Physical Constants

Physical Constants

Selected for Students

W. H. J. CHILDS

D.Sc., Ph.D., F.Inst.P., F.R.S.E.

Emeritus Professor of Physics,
Heriot-Watt University,
Edinburgh.

CHAPMAN AND HALL
and
SCIENCE PAPERBACKS

First published 1934
by Methuen & Co Ltd
Ninth edition 1972
published by Chapman and Hall Ltd
11 New Fetter Lane London EC4P 4EE

Reprinted 1980

© *1972 W.H.J. Childs*

ISBN-13: 978-0-412-21050-1   e-ISBN-13: 978-94-011-6930-1
DOI: 10.1007/978-94-011-6930-1

Preface

Time changes most things, and physical constants are seem-
ingly not immune from this contingency. The publishers'
suggestion for a new edition has provided an opportunity for
a thorough scrutiny of the contents of this book and for the
adjustment to some fourteen years of changes since the last
edition. At the same time the contents have been increased by
the addition of one or two new tables, and current moves
towards the adoption of an international system of units ex-
plain the appearance in the book, wherever it is practicable,
of S.I. units. However, the habits of a lifetime are not easily
changed, and it is too much to expect that despite careful
proofreading the change of units has been carried through
without errors. I shall be grateful for intimation of such mis-
takes. I wish moreover to thank all those who in the past have
taken the trouble to draw my attention to errors, or to superior
determinations, or simply to offer advice or criticism.

For those who do not find what they seek in this collection,
more complete compilations of physical constants are to be
found in:

Tables of Physical and Chemical Constants, G.W.C. Kaye
and T.H. Laby, 13th edition 1966, Longman.
Handbook of Chemistry and Physics, 52nd edition 1971,
Chemical Rubber Company.
International Critical Tables. National Research Council,
McGraw-Hill. *Zahlenwerte und Funktionen aus Physik*;
Chemie; *Astronomie*; *Geophysik and Technik*. Landolt and
Bornstein, Springer-verlag.
Extensive tables of spectroscopic wavelength measure-
ments are published in:
Tabelle der Hauptlinien aller Elemente. H. Kayser. 1926,
Springer.

Wavelength Tables. G.R. Harrison, 1939, Mass. Inst. Tech., New York.

I am indebted to B.N. Taylor, W.H. Parker and D.N. Langenberg for permission to quote the values for most of the general fundamental constants shown on Pages 91 and 93 of this book (and used indirectly elsewhere) and which they recommend as a result of their critical analysis of current theoretical and experimental developments (*Rev. Mod. Phys.* and Academic Press).

W. H. J. C.

HERIOT-WATT UNIVERSITY,
EDINBURGH.
February, 1972.

Contents

7

CONTENTS

8

CONTENTS

Units

In course of time the system of units based on the metre, kilogram, second and ampere (MKSA system), has found increasing international acceptance. This system, with extension to cover measurements in the fields of thermodynamics, photometry and chemistry, was put forward in 1960 as the Système International d'Unités (SI) and promises eventually to become the legal system in all metric countries.

Basic Units

Length (l). The metre (m) is not now based on the platinum—iridium prototype kept at Paris, but is defined as the length equal to 1 650 763.73 wavelengths in vacuum of the radiation corresponding to the transition between the levels $2p_{10}$ and $5d_5$ of the krypton-86 atom.

Mass (m). The kilogram (kg) is, as heretofore, the mass of the platinum—iridium prototype kept at Sèvres. It was originally based on the metre and was intended to be the mass of a cubic decimetre of pure water at 4°C.

Time (t). The second (s) is no longer based on the mean solar day, or on the tropical year for 1900 January 0, but is now defined as the duration of 9 192 631 770 periods of the radiation corresponding to the transition between the two hyperfine levels of the ground state of the caesium-133 atom.

Electric current (I). The ampere (A); the constant current which produces a force of 2×10^{-7} newtons per metre between two parallel conductors through which it flows, if these conductors are infinitely long, of negligible cross section and spaced in vacuo 1 m apart.

11

PHYSICAL CONSTANTS

Thermodynamic temperature (*T*). The unit is the kelvin, (K) the fraction 1/273.16 of thermodynamic temperature of the triple point of water.

Luminous intensity (*I*). The candela (cd) replaces units based on flame or filament standards. It is the luminous intensity, in the perpendicular direction, of a surface of 1/600 000 square metre of a black body at the temperature of freezing platinum under a pressure of 101 325 newton per square metre (standard atmospheric pressure).

Amount of substance (ν, *n*). The mole (mol) is the amount of substance of a system which contains as many elementary entities as there are atoms in 0.012 kilogram of carbon-12. The elem ntary entities must be specified and may be atoms, molecules ions, electrons, other particles or specified groups of particles.

These seven un s comprise the basis of SI and from them are formed the variou derived units. Decimal multiples and sub-multiples are expressed by prefixes as follows:

Multiples: 10^1 deca (da), 10^2 hecto (h), 10^3 kilo (k), 10^6 mega (M), 10^9 giga (G), 10^{12} tera (T).
Submultiples: 10^{-1} deci (d), 10^{-2} centi (c), 10^{-3} milli (m), 10^{-6} micro (μ), 10^{-9} nano (n), 10^{-12} pico (p), 10^{-15} femto (f), 10^{-18} atto (a).

Derived units

Length. Kilometre (km), centimetre (cm), millimetre (mm). The name 'micron' is abandoned, and the unit becomes the micro-metre (μm) = 10^{-6} m. The angstrom (Å) = 10^{-10} m is to be retain-ed for a limited time, but it is recommended that the nanometre (nm) should be used.

Square measure (l^2). Square metre (m²), square centimetre (cm²). The are (a) = 10^2m², and the hectare (ha) = 10^4m² are to be retained for a limited time.

Volume (l^3). Cubic metre (m³). The litre (l) = 1 dm³ is retained, but is divorced from any connection with a 1 kg mass of pure water and reverts to its original meaning. It is recommended that the name 'litre' shall no longer be used to express volume measurements of high accuracy.

12

UNITS

Mass. Gram (g), decigram (dg), centigram (cg), milligram (mg). The practice of using these as units of force (i.e. weight) for example kilogram-force (kgf) is deprecated. Instead it is recommended to use the SI unit of force (newton). $1\,\text{kgf} = 9.806\,65$ newton.

Supplementary units: Units of plane and solid angle.

Plane angle. The radian (rad) is the plane angle subtended at the centre of a circle by an arc of its circumference equal in length to the radius. $(1\,\text{rad} = 57.296°.)$

Solid angle. The steradian (sr) is the solid angle subtended at the centre of a sphere by an area of its surface equal in magnitude to the square of its radius.

General definitions and units

Velocity (v; $l\,t^{-1}$). A point which moves in a straight line with uniform rate of change of position is said to have uniform velocity. The unit is a rate of change of position of 1 metre per second. Rate of change of position irrespective of direction (i.e. a scalar quantity) is denoted by the term speed. The terms speed and velocity are often used without distinction.

Angular velocity (ω; $\text{rad}\,t^{-1}$), or rate of change of angle. A rate of change of 1 radian per second is the unit.

Acceleration (a; $l\,t^{-2}$). or rate of change of velocity. When the velocity of a point changes at the rate of 1 metre per second per second it is said to experience unit acceleration.

Angular acceleration (α; $\text{rad}\,t^{-2}$), or rate of change of angular velocity. The unit is 1 radian per second per second.

Force (F; $l\,m\,t^{-2}$). A mass which moves with acceleration is said to be acted on by a force, which is defined to be the product mass × acceleration. The unit is the newton (N), that force which, acting on a mass of one kilogram, produces an acceleration of 1 metre per second per second.

Energy or Work (W; $l^{2}\,m\,t^{-2}$). When a force moves its point of application in the direction in which the force is acting it is said to do work. The work done in this way by a force of 1 newton moving over a distance of 1 metre is the unit of work, the joule (J).

Power (P; $l^2 m t^{-3}$), or the rate at which work is done. The unit is a rate of 1 joule per second, the watt (W).

Density (ρ; $m\, l^{-3}$), or the mass of unit volume of a substance. A substance of which 1 cubic metre has a mass of 1 kilogram is of unit density. Note that in SI units the density of water is 1000 kg m^{-3}. The relative density ('specific gravity') of a substance is the ratio of its density to that of water.

Pressure (p; $l^{-1} m t^{-2}$). When a force is transmitted to act over an area it is said to exert a pressure; unit pressure is defined as unit force per unit area, in SI the pascal (Pa), or 1 newton per square metre (N m^{-2}). The meteorological unit, the bar (10^5 N m^{-2}), and the standard atmosphere (atm), $101\,325 \text{ N m}^{-2}$ (0°C, 760 mmHg, $g = 9.806\,65 \text{ m s}^{-2}$) are retained for a limited time.

Frequency (ν; f). The unit is the hertz (Hz), one cycle per second. (s^{-1}).

Electrical and Magnetic Units

Current (I). The ampere is a basic unit and has already been defined (p. 11).

Charge (Q). The unit is the total charge transferred when a current of 1 ampere flows for 1 second, the coulomb (C). Thus in terms of basic units 1 coulomb is 1 s A.

Potential (V). The unit is the volt (V). A difference of potential of 1 volt exists between two points if 1 joule of work is done when 1 coulomb is transferred between them. One volt equals $1 \text{ m}^2 \text{ kg s}^{-3} \text{A}^{-1}$.

Resistance (R). The resistance of a conductor which passes 1 ampere when a difference of potential of 1 volt exists between its ends is 1 ohm (Ω). In basic units 1 ohm equals $1 \text{ m}^2 \text{ kg s}^{-3} \text{ A}^{-2}$.

Electric field strength (E). The unit, one volt per metre, exists at a point if the force exerted on 1 coulomb at the point is 1 newton. In basic units 1 V m^{-1} equals $1 \text{ m kg s}^{-3} \text{ A}^{-1}$.

Electric displacement or flux density (D), is the sum of the vectors polarization (P) and the product of the field strength (E) and the permittivity (ϵ). In an isotropic medium these vectors are collinear and in vacuo $D = \epsilon_0 E$. The unit is 1 coulomb per m². In basic units (m^{-2} s A).

Capacitance (C). The unit, the farad (F) or coulomb per volt, is possessed by a conductor if a charge increase of one coulomb raises its potential by one volt. In basic units one farad equals $1\,m^{-2}\,kg^{-1}\,s^4\,A^2$.

Magnetic field strength (H). The unit, 1 ampere per metre (A m^{-1}) exists at the centre of a one-turn circular coil of diameter 1 metre, carrying a current of 1 ampere.

Magnetic induction or flux density (B), is the sum of the vectors intensity of magnetization (M, J) and the product of magnetic field strength (H) and permeability (μ). In an isotropic medium these vectors are collinear, and in vacuo $B = \mu_0 H$. The unit is the tesla (T), or 1 weber per m^2 (Wb m^{-2}), in basic units $kg\,s^{-2}\,A^{-1}$.

Magnetic flux (Φ). The magnetic flux threading an area ds, sufficiently small to render the flux density (B) uniform over it, is defined to be the scalar product $B \cdot ds$. The flux threading a finite area S is then $\Phi = \int B \cdot ds$. The unit is the weber (Wb) or volt-second ($m^2\,kg\,s^{-2}\,A^{-1}$).

Self inductance (L). The unit is possessed by a circuit in which unit current causes a total flux of 1 weber to thread the circuit. The unit is the henry (H), or weber per ampere, in basic units $m^2\,kg\,s^{-2}\,A^{-2}$.

Mutual inductance (M). The unit is possessed by two circuits if unit current in either causes a total flux of 1 weber to thread the other. The unit is the henry (H).

With the adoption, in SI, of the particular units of force and charge, it is necessary to introduce a constant μ_0, the permeability of free space. Then since $(\mu_0 \epsilon_0)^{-\frac{1}{2}} = c$, the velocity of light in free space, and further $(\mu_0/\epsilon_0)^{\frac{1}{2}} = z_0$, the permittivity ϵ_0 and the impedance z_0 of free space are likewise fixed. We have then $\mu_0 = 4\pi\,10^{-7}\,H\,m^{-1} = 1.256\,637 \times 10^{-6}\,H\,m^{-1}$. $\epsilon_0 = 8.854\,16 \times 10^{-12}\,F\,m^{-1}$; $z_0 = 376.371\,\Omega$.

Heat Units

Quantity of heat (Q). The unit is the joule (J). The continued use of the various calorie (cal) units, e.g. the 15°C calorie (4.1855 J), International Steam Table calorie (4.1868 J) is deprecated.

Temperature (T). The thermodynamic temperature unit, the kelvin (K) is a basic unit and has already been defined (p. 12). In addition, Celsius temperature (t) is also permissible, with $t = T - T_0$, where $T_0 = 273.15$ K. Thus the unit degree Celsius is also the unit kelvin.

THE CENTIMETRE, GRAM, SECOND (CGS) SYSTEM OF UNITS
(SI equivalent in parentheses)

The older system of units based on the centimetre, gram and second is likely to find decreasing use with the passage of time. Some of the units of this system are:

Volume. The practical unit is the millilitre. The litre (i.e. the volume of 1 kg of pure water at 4°C) was originally intended to be 1000 cm³, but is now recognised to be 1000.028 cm³. Thus 1 cm³ = 0.999 972 millilitre. Note that the definition of the litre in the CGS system is different from that of SI. It is recommended that the name 'litre' shall no longer be used to express volume measurements of high accuracy.

Force. The unit is the dyne, that force which acting on a mass of 1 gram, produces an acceleration of 1 centimetre per second per second. (1 dyne = 10^{-5} N)

Energy or work. When a force of 1 dyne moves its point of application 1 centimetre in the direction in which the force is acting it performs the unit of work, 1 erg. (1 erg = 10^{-7} J)

Power. The unit is a rate of working of 1 erg per second. 1 erg per s = 10^{-7} W. 1 horse-power (hp) = 746 W.

Density. The mass of unit volume of a substance. A substance of which 1 millilitre has a mass of 1 gram is of unity density.

Electrical and Magnetic units

Charge. The electromagnetic unit, e.m.u., is the quantity delivered by unit current in one second (10 C). The electrostatic unit, e.s.u. repels an equal quantity 1 cm away in vacuo with a force of 1 dyne (10/c or approx. $10^{-9}/3$ C).

Current. 1 e.m.u. flowing in a circle of radius 1 cm exerts a force of 2π dynes on a unit pole at the centre (10 A). The e.s.u. is a rate of flow equal to unit charge per second (10/c or approx. $10^{-9}/3$ A).

16

Electromotive force or potential difference. The e.m.u. confers on 1 e.m.u. charge the ability to do 1 erg of work (10^{-8}V).
The e.s.u. confers on 1 e.s.u. charge the ability to do 1 erg
of work (300 V).

Resistance. Allows unit e.m.f. to produce unit current.
1 e.m.u. = $10^{-9}\Omega$, 1 e.s.u. = 10^{-9} c² (approx. 9×10^{11} Ω)

Capacitance. The unit is possessed by a conductor when an
increase of unit charge will raise its potential by unity.
1 e.m.u. = 10^{9}F, 1 e.s.u. = 10^{9}c^{-2} (approx. 10^{-11}/9 F) A sphere
of radius 1 cm has in vacuo a capacitance of 1 e.s.u.

Magnetic pole strength. 1 e.m.u. magnetic pole repels an equal
pole 1 cm away in vacuo with a force of 1 dyne.

Magnetic field strength. 1 e.m.u. is possessed by a field which
exerts a force of 1 dyne on unit pole. (1 oersted = $10^{3}/4\pi$
= 79.577 A m^{-1} , 1 e.s.u. = $10^{3}/(4\pi c)$ = 2.6544 10^{-9} A m^{-1})

Self inductance. Unit self inductance is possessed by a circuit
in which unit electromotive force is induced by unit rate of
change of current. (1 e.m.u. = 10^{-9}H; 1 e.s.u. = 10^{-9} c^2, approx.
9×10^{11} H) Also, a circuit in which unit current causes a total
magnetic flux of 1 maxwell (10^{-8} Wb) to thread the circuit.

Mutual inductance. Unit mutual inductance is possessed by
two circuits if a rate of change of current of one unit per second in either induces unit electromotive force in the other.
(1 e.m.u. = 10^{-9} H; 1 e.s.u. = 10^{-9} c^2, approx. 9×10^{11} H) Also,
if unit current in either causes a total flux of 1 maxwell
(10^{-8} Wb) to thread the other.

  (In the above expressions, c = velocity of light in vacuo.)

ASTRONOMICAL AND GEODETICAL DATA

POSITIONS OF SELECTED FIXED STARS

1 January 1972

Name	Distance in light-years	Right ascension h	min	s	Declination o	
α Ursae Minor (Pole star)	460	2	11	55	+89	10
α Tauri (Aldebaran)	57	4	34	46	+16	28
β Orionis (Rigel)	540	5	13	33	− 8	14
α Aurigae (Capella)	43	5	15	10	+45	59
α Canis Majoris (Sirius)	8.8	6	44	15	−16	41
α Leonis (Regulus)	56	10	07	22	+12	04
α Virginis (Spica)	360	13	24	09	−11	03
α Bootis (Arcturus)	40	14	14	45	+19	17
α Scorpii (Antares)	125	16	28	09	−26	23
α Lyrae (Vega)	27	18	36	14	+38	46
α Aquilae (Altair)	16	19	49	51	+ 8	49
α Pegasi (Markab)	85	23	03	45	+15	06

1 light year $= 9.46 \times 10^{12}$ km
1 parsec (pc) $= 30.86 \times 10^{12}$ km

MISCELLANEOUS CONSTANTS

$1°$ of latitude at the poles $=$ 111.71 km
$1°$ of latitude at the equator $=$ 110.56 km
Constant of stellar aberration $=$ $20.\underline{496}''$
Constant of Earth's precession $=$ $50.\underline{26}''$ per year
Nutation $=$ $9.21''$
Gravitational constant (G) $=$ $6.670 \ 10^{-11}$ m^3 kg^{-1} s^{-2}
Length of seconds pendulum (period $2s$) $=$ g/π^2
at latitude $45°$ $=$ 0.99357 m
Formula for calculating g at any place:

$g = 9.806 \ 15 - 0.025 \ 862 \cos(2 \times \text{latitude}) + 5.9 \times 10^{-5}$
$\cos^2(2 \times \text{latitude}) - 3 \times 10^{-6}$ (ht. in metres)

18

Volume per cent at sea level		Volume per cent at sea level	
78.09	Nitrogen	0.8×10^{-5}	Xenon
20.95	Oxygen	1.8×10^{-3}	Neon
0.93	Argon	5.24×10^{-4}	Helium
0.03	Carbon dioxide	5.0×10^{-5}	Hydrogen
10^{-4}	Krypton		

Stratosphere begins at approx. 12 km.
Equivalent height of the Heaviside–Kennelly layer (Region E), 100 km.
Equivalent height of the Appleton layer (Region F), 230 km.
Mean temperature gradient = 6.5 K per km.
Voltage gradient, average value in fine weather = 100 V m^{-1}

THE EARTH

Approximate age of Earth 4.5×10^{9} y
Area of land surface 57.49×10^{6} (mile)^{2}; 148.90×10^{12} m²
Area of water surface 139.46×10^{6} (mile)^{2}; 361.20×10^{12} m²
Highest mountain (Everest) 8847.7 m
Greatest ocean depth (Marianas Trench) 11 033.2 m
Ocean mass 1.42×10^{21} kg
Mass of atmosphere 5.27×10^{18} kg
Velocity of seismic waves (within 200 km depth):
    Compressive wave 8 km s^{-1}; shear wave 4.5 km s^{-1}

THE PLANETS

Name	Mean dist. from sun, 10⁶ km	Period	Period of axial rotation	Eccent. of orbit	Inclin. of orbit to ecliptic	Equatorial diameter, km	Oblateness	Mass, 10²⁴ kg m	Density 10³ kg m⁻³	No. of satellites	"g" m s⁻²	Inclin. of equator to orbit
Mercury	57.85	87.97 d	59 d	0.206	7 0	4 840	0	0.329	5.54	0	3.75	0
Venus	108.11	224.70 d	244 d	0.007	3 24	12 300	0	4.9	5.03	0	8.64	175°
Earth	149.46	365.26 d	23h 55 m	0.017	0 0	12 756.6	1/298.2	6.0	5.53	1	9.806 15	23° 27'
Mars	227.7	686.98 d 1 y 322 d	24h 37 m 23 s	0.093	1 51	6 783	1/192	0.65	3.98	2	3.77	23° 30'
Jupiter	777.6	11 y 314 d	9h 50 m	0.048	1 18	142 900	1/15	1899.8	1.34	12	24.82	3° 7'
Saturn	1426	29 y 167 d	10h 14 m	0.056	2 29	119 500	1/9.5	568.8	0.71	9 and 3 rings	10.63	26° 45'
Uranus	2868.3	84 y 5 d	10h 45 m	0.047	0 46	47 200	1/14	87.7	1.67	5	10.50	98°
Neptune	4494.3	164 y 288 d	15h 48 m	0.009	1 47	44 650	1/40	104	2.3	2	13.90	28° 48'
Pluto	5900	247 y 255 d	—	0.253	17 09	—	—	—	—	—	—	—

CONSTANTS OF THE SUN, EARTH AND MOON

Name	Mean angular diameter	Diameter, km	Parallax	Mass, kg	Density	g, m s^{-2}	Temp. °C	Period of axial rotation	Other constants
Sun	32′ 6.5″	1.392×10^5	8.79″	1.990×10^{30}	1.39	271×10^3	approx. 6000	25 d 9.1 h	Solar constant 1400 ± 30 W m^{-2}
Earth	—	equator 12 756.3 polar 12 713.6	—	5.976×10^{24}	5.517	9.806 15 (lat. 45°)	—	23 h 56 min 4.1 s	Distance to sun $= 149.67 \times 10^6$ km Distance to moon $= 384.4 \times 10^3$ km Sidereal year $= 365$ d 6 h 9 min 10 s Inclin. of equator to ecliptic $= 23° \, 26′ \, 35.0″$
Moon	31′ 6″	3478	57′ 2.7″	7.35×10^{22}	3.39	1.62	approx. 120 (lunar day)	27 d 7 h 43 min 11 s	Sidereal month $= 27$ d 7 h 43 min 11 s

12 o'clock midday (12 GMT) at Greenwich; each country is given with its corresponding time according to the 24-hour system. Countries which use summer time are printed in italics.

The Commonwealth etc.

Australia
  Capital Territory 22
  Northern Territory 21½
  New South Wales 22
  Queensland 22
  South Australia 21½
  Victoria 22
  Tasmania 22
  Western Australia 20
Bangladesh 18
Barbados 8
Bermudas 8
Botswana 13½
Canada
  E. of long. 68° 8
  Long. 68°–85°
  (in north) 7
  Long. 68°–90°
  (in south) 7
  Long. 85°–102°
  (in north) 6
  Long. 90°–102°
  (in south) 6
  W. of long. 102° 5
Ceylon 17½
Cyprus 14
Falkland Is. 8
Fiji 24
Gambia 12
Ghana 12
Gibraltar 13
Guyana 8¼
Hongkong 20
India 17½
Jamaica 7
Kenya 15
Lesotho 13½
Malawi 13½
Malaysia 20
Malta G.C. 13
Mauritius 16
Nauru 23½
New Zealand 24
Nigeria 13
Pakistan 17
Seychelles 16
Sierra Leone 12
Singapore 19½
Swaziland 13½
Tanzania 15
Tonga 1 †

Trinidad & Tobago 8
Uganda 15
United Kingdom 12
Western Samoa 1
Zambia 13½

Principal Foreign Countries

Albania 13
Algeria 12, 13
America, United States
  Eastern Time 7
  Central Time 6
  Mountain Time 5
  Pacific Time 4
  Alaska 1, 2, 3, 4
Argentina 9
Austria 13
Belgium 13
Bolivia 8
Brazil
  East 9
  Central 8
  West 7
Bulgaria 14
Burma 18½
Cambodia 19
Cameroon Rep. 13
Central African Rep. 13
Chile 8
China E. 20
Colombia 7
Congo Rep. 13
Cuba 7
Czechoslovakia 13
Denmark 13
Dahomey 17
Eire 12
Ethiopia 15
Ecuador 7
Finland 14
Formosa 20
France 13
Germany 13
Greece 14
Guatemala 6
Hawaii 2
Honduras 6
Hungary 13

Iceland 12
Indonesia 21
Iran 15½
Iraq 15
Israel 14
Italy 13
Ivory Coast 12
Japan 21
Jordan 14
Korea 21
Labrador 8½
Laos 19
Lebanon 13½
Libya 13½
Luxembourg 13
Mauritania 12
Mali 12
Mexico 4, 5, 6
Morocco 12
Netherlands 13
Nicaragua 6
Nigeria 12
Norway 13
Panama 7
Paraguay 8
Peru 7
Poland 13
Portugal 13
Rhodesia 14
Rumania 14
Somalia 15
South Africa 13½
Spain 13
Sudan 13½
Sweden 13
Switzerland 13
Syria 14
Thailand 19
Tunisia 13
Turkey 14
U.S.S.R.
  40°–52° 30'E 16
  W. of 52°30' 15
United Arab Rep. 1
Uruguay 8½
Venezuela 8
Vietnam
  South 20
  North 19
Yugoslavia 13
Zaire 13

† on the suceeding day

22

GENERAL PHYSICS

THE REDUCTION OF WEIGHTS TO VACUO

1 apparent gram of a substance weighed in air has a real weight of

$$1 + \rho_{(air)}[1/\rho_{(subs)} - 1/\rho_{(weights)}]_{gram}$$

Density of dry air at $20°C$ and $760\,mmHg$ pressure is $1.205\,kg\,m^{-3}$.

Density of brass weights 8.40, aluminium weights $2.70 \times 10^3\,kg\,m^{-3}$.

See Nomogram 1.

MOMENTS OF INERTIA (*l*)

In each case the axis of rotation, unless otherwise stated, passes through the centre of mass of the body. *m* is the total mass.

Body	Moment of inertia
Massive point, distant *r* from the axis of rotation	mr^2
Uniform thin rod of length *l*, in the plane of rotation	$ml^2/12$
The same, but axis at one end . . .	$ml^2/3$
Rectangular lamina ($a \times b$) in the plane of rotation	$m(a^2 + b^2)/12$
Rectangular lamina ($a \times b$) with *a* in, and *b* perpendicular to the plane of rotation .	$ma^2/12$
Rectangular parallelopiped ($a \times b \times c$) with the plane *a, b* in the plane of rotation	$m(a^2 + b^2)/12$
Circular disc, radius *r*, in the plane of rotation	$mr^2/2$
Circular disc, radius *r*, rotating about a diameter	$mr^2/4$
Cylinder of radius *r*, length *l*, axis in the plane of rotation	$m\left(\dfrac{r^2}{4} + \dfrac{l^2}{12}\right)$
Sphere of radius *r*	$2mr^2/5$
Anchor ring, mean radius *R*, cross-sectional radius *r*	$m(R^2 + 3r^2/4)$

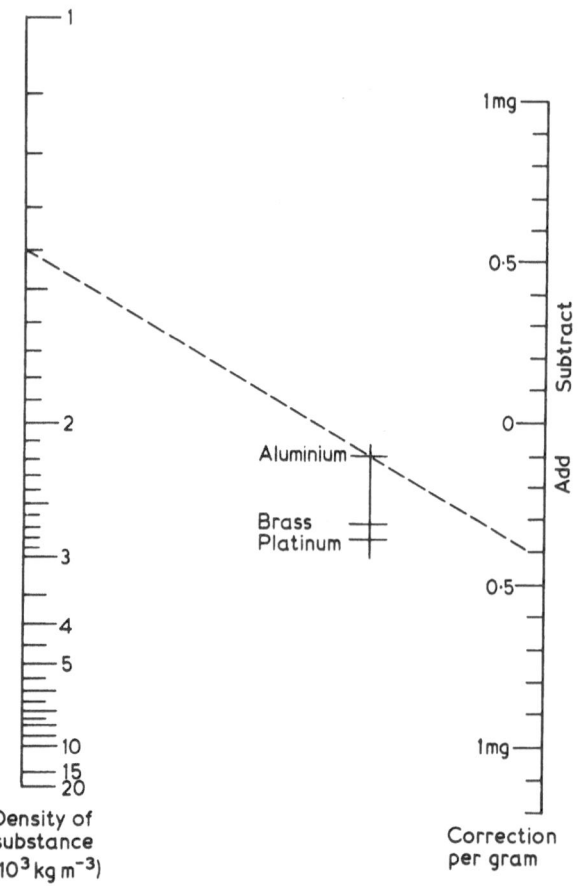

Nomogram 1. The reduction of weights to vacuo.

A straight edge placed across the diagram passes through three inter-related quantities. Thus for a substance of density 1.4×10^3 weighed with aluminium weights, the correction is 0.42 mg to be added to every 1 g apparent weight.

24

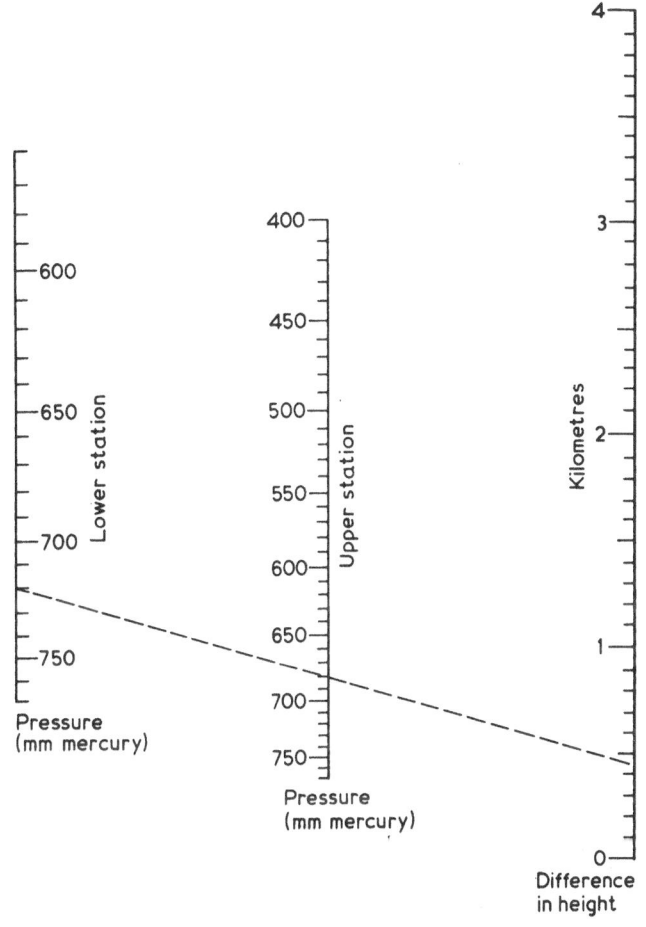

Nomogram 2. Height from barometer readings.

A straight edge placed across the diagram passes through three inter-related quantities. For example, if the barometer reads 720 mm at the lower station and 680 mm at the upper station, the vertical distance between the two stations is 0.45 km.

25

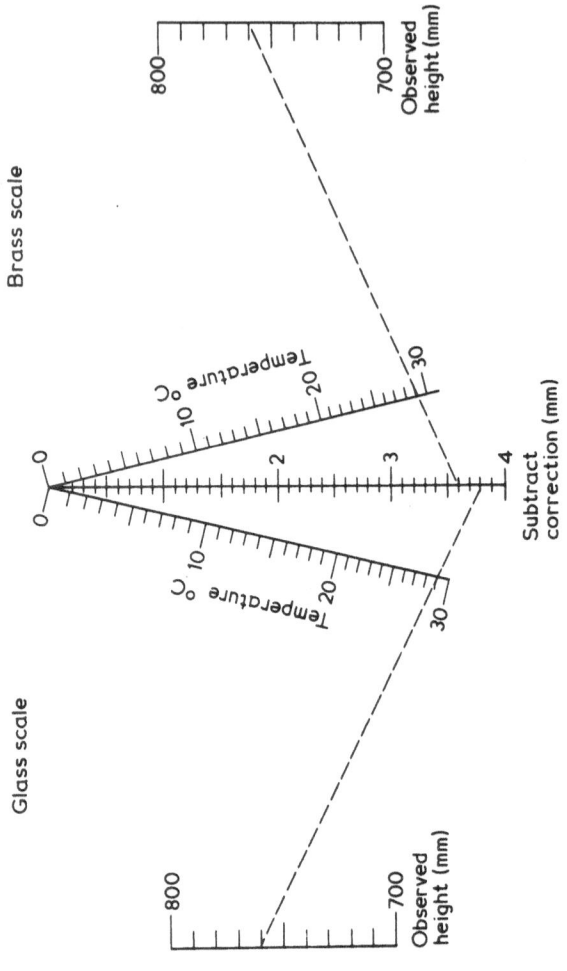

Nomogram 3. Barometer temperature correction, to 0°C.
A straight edge placed across either half of the diagram passes
through three inter-related quantities. Thus if the barometer reads
760 mm and the temperature is 29°C, the correction to be subtracted
is 3.81 mm if the barometer has a glass scale, 3.58 mm if it has a
brass scale.

26

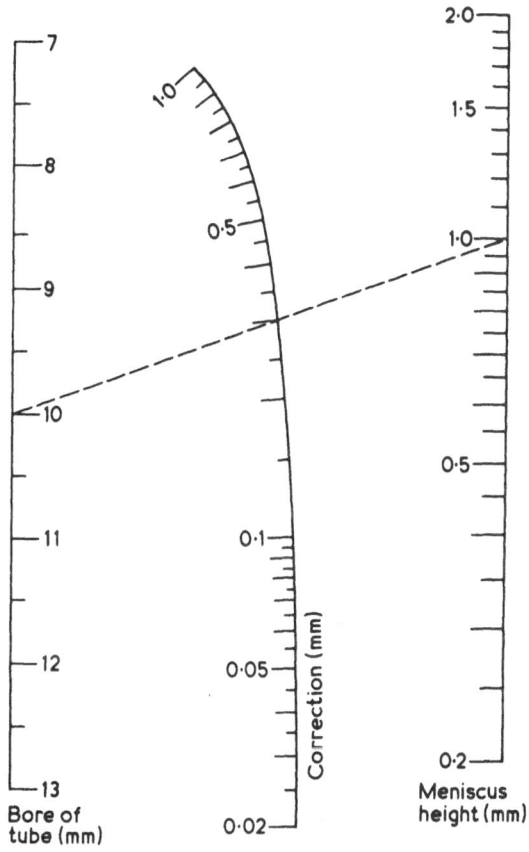

7

8

9

10

11

12

13

Bore of
tube (mm)

1·0

0·5

0·1

0·05

0·02

Correction (mm)

2·0

1·5

1·0

0·5

0·2

Meniscus
height (mm)

Nomogram 4. Meniscus correction for mercury columns.

A straight edge placed across the diagram passes through three inter-related quantities. Read to the top of the meniscus and *add* the correction. Example: For a 1 cm diameter tube and a meniscus height of 1 mm the correction to be added is 0.3 mm.

ATOMIC WEIGHTS (1969)

The atomic number is adjacent to each name.
The atomic weight 12.0000 is assumed for the ^{12}C isotope.

Symbol	No.	Name	Weight	Symbol	No.	Name	Weight
Ag	47	Silver	107.868	Mo	42	Molybdenum	95.94
Al	13	Aluminium	26.981 54	N	7	Nitrogen	14.006 7
Ar	18	Argon	39.948	Na	11	Sodium	22.989 77
As	33	Arsenic	74.921 6	Nb	41	Niobium	92.906 4
Au	79	Gold	196.966 5	Nd	60	Neodymium	144.24
B	5	Boron	10.81	Ne	10	Neon	20.179
Ba	56	Barium	137.33	Ni	28	Nickel	58.70
Be	4	Beryllium	9.012 18	O	8	Oxygen	15.999 4
Bi	83	Bismuth	208.980 4	Os	76	Osmium	190.2
Br	35	Bromine	79.904	P	15	Phosphorus	30.973 76
C	6	Carbon	12.011	Pb	82	Lead	207.2
Ca	20	Calcium	40.08	Pd	46	Palladium	106.4
Cd	48	Cadmium	112.41	Pr	60	Praseodymium	140.907 7
Ce	58	Cerium	140.12	Pt	78	Platinum	195.09
Cl	17	Chlorine	35.453	Ra	88	Radium	226.025 4
Co	27	Cobalt	58.933 2	Rb	37	Rubidium	85.467 8
Cr	24	Chromium	51.996	Re	75	Rhenium	186.207
Cs	55	Caesium	132.905 4	Rh	45	Rhodium	102.905 5
Cu	29	Copper	63.546	Ru	44	Ruthenium	101.07
Dy	66	Dysprosium	162.50	S	16	Sulphur	32.06
Er	68	Erbium	167.26	Sb	51	Antimony	121.75
Eu	63	Europium	151.96	Sc	21	Scandium	44.955 9
F	9	Fluorine	18.998 403	Se	34	Selenium	78.96
Fe	26	Iron	55.847	Si	14	Silicon	28.085 5
Ga	31	Gallium	69.72	Sm	62	Samarium	150.4
Gd	64	Gadolinium	157.25	Sn	50	Tin	118.69
Ge	32	Germanium	72.59	Sr	38	Strontium	87.62
H	1	Hydrogen	1.007 9	Ta	73	Tantalum	180.947 9
He	2	Helium	4.002 60	Tb	65	Terbium	158.925 4
Hf	72	Hafnium	178.49	Te	52	Tellurium	127.60
Hg	80	Mercury	200.59	Th	90	Thorium	232.038 1
Ho	67	Holmium	164.930 4	Ti	22	Titanium	47.90
I	53	Iodine	126.904.5	Tl	81	Thallium	204.37
In	49	Indium	114.82	Tm	69	Thulium	168.934 2
Ir	77	Iridium	192.22	U	92	Uranium	238.029
K	19	Potassium	39.098 3	V	23	Vanadium	50.941 5
Kr	36	Krypton	83.80	W	74	Tungsten	183.85
La	57	Lanthanum	138.905 5	X	54	Xenon	131.30
Li	3	Lithium	6.941	Y	39	Yttrium	88.905 9
Lu	71	Lutetium	174.967	Yb	70	Ytterbium	173.04
Mg	12	Magnesium	24.305	Zn	30	Zinc	65.37
Mn	25	Manganese	54.938 0	Zr	40	Zirconium	91.22

PROPERTIES OF SEMICONDUCTOR ELEMENTS

Element	m.p. °C	b.p. °C	Density 10^3 kg m^{-3}	Youngs mod. E 10^{10} N m^{-2}	Bulk mod. K 10^{10} N m^{-2}	Shear mod. G 10^{10} N m^{-2}	Linear expans. coeff. 10^{-6} °C^{-1}	Sp.ht. capacity J kg^{-1} K^{-1}	Thermal cond. W m^{-1} K^{-1}	Resist. (20°C) 10^{-8} Ω m	Magn. Suscept. 10^{-9} kg^{-1}
14 Silicon	1410	2355	2.325	11.26	10.1	—	2.5	706	175	$\sim 10^{10}$	− 0.17
31 Gallium	29.78	2403	5.878 (29.6°C) / 6.068 (29.8°C)	—	—	—	19.2	377	34	17.4	− 0.38
32 Germanium	937	2830	5.307 (25°C)	—	7.9	—	5.7	324	59	$46 . 10^6$	− 0.13
33 Arsenic	613 (subl.)	–	5.73	—	—	—	6	335	—	33.3	− 0.39
34 Selenium	217	685	4.79	—	—	—	26	324	0.24	$\sim 10^{12}$	− 0.40
51 Antimony	630.74	1750	6.679 (20 °C)	7.78	—	1.98	11	210	18.5	44	− 1.02
52 Tellurium	452	1390	6.2	—	—	—	17	201	50	$16 . 10^4$	− 0.39
83 Bismuth	271.3	1560	9.730 (20°C)	3.19	3.13	1.20	13.5	112	8.7	117	− 1.68

ELECTRON-SHELL STRUCTURE, GROUND STATE AND IONIZATION POTENTIAL OF NEUTRAL ATOMS OF THE ELEMENTS

(K) represents a completed K shell $1s^2$

(L) represents a completed L shell $2s^2 2p^6$

($\underline{M}$) represents a completed M shell $3s^2 3p^6 3d^{10}$

($\overline{N}$) represents an incomplete N shell $4s^2 4p^6 4d^{10}$

(N) represents a complete N shell $4s^2 4p^6 4d^{10} 4f^{14}$

Element	Structure	Ground state	Ionization potential (eV)
1 H	$1s^1$	$^2S_{1/2}$	13.60
2 He	$1s^2$	$^1S_0$	24.58
3 Li	(K) $2s^1$	$^2S_{1/2}$	5.39
4 Be	(K) $2s^2$	$^1S_0$	9.32
5 B	(K) $2s^2 2p^1$	$^2P_{1/2}$	8.30
6 C	(K) $2s^2 2p^2$	$^3P_0$	11.26
7 N	(K) $2s^2 2p^3$	$^4S_{3/2}$	14.53
8 O	(K) $2s^2 2p^4$	$^3P_2$	13.61
9 F	(K) $2s^2 2p^5$	$^2P_{3/2}$	17.42
10 Ne	(K,L)	$^1S_0$	21.56
11 Na	(K,L) $3s^1$	$^2S_{1/2}$	5.14
12 Mg	(K,L) $3s^2$	$^1S_0$	7.64
13 Al	(K,L) $3s^2 3p^1$	$^2P_{1/2}$	5.98
14 Si	(K,L) $3s^2 3p^2$	$^3P_0$	8.15
15 P	(K,L) $3s^2 3p^3$	$^4S_{3/2}$	10.48
16 S	(K,L) $3s^2 3p^4$	$^3P_2$	10.36
17 Cl	(K,L) $3s^2 3p^5$	$^2P_{3/2}$	13.01
18 A	(K,L) $3s^2 3p^6$	$^1S_0$	15.76
19 K	(K,L) $3s^2 3p^6 4s^1$	$^2S_{1/2}$	4.34
20 Ca	(K,L) $3s^2 3p^6 4s^2$	$^1S_0$	6.11
21 Sc	(K,L) $3s^2 3p^6 3d^1 4s^2$	$^2D_{3/2}$	6.54
22 Ti	(K,L) $3s^2 3p^6 3d^2 4s^2$	$^3F_2$	6.82
23 V	(K,L) $3s^2 3p^6 3d^3 4s^2$	$^4F_{3/2}$	6.74
24 Cr	(K,L) $3s^2 3p^6 3d^5 4s^1$	$^7S_3$	6.76
25 Mn	(K,L) $3s^2 3p^6 3d^5 4s^2$	$^6S_{3/2}$	7.43
26 Fe	(K,L) $3s^2 3p^6 3d^6 4s^2$	$^5D_4$	7.87
27 Co	(K,L) $3s^2 3p^6 3d^7 4s^2$	$^4F_{3/2}$	7.86
28 Ni	(K,L) $3s^2 3p^6 3d^8 4s^2$	$^3F_4$	7.63
29 Cu	(K,L,M) $4s^1$	$^2S_{1/2}$	7.72
30 Zn	(K,L,M) $4s^2$	$^1S_0$	9.39
31 Ga	(K,L,M) $4s^2 4p^1$	$^2P_{1/2}$	6.06
32 Ge	(K,L,M) $4s^2 4p^2$	$^3P_0$	7.88
33 As	(K,L,M) $4s^2 4p^3$	$^4S_{3/2}$	9.81
34 Se	(K,L,M) $4s^2 4p^4$	$^3P_2$	9.75
35 Br	(K,L,M) $4s^2 4p^5$	$^2P_{3/2}$	11.84
36 Kr	(K,L,M) $4s^2 4p^6$	$^1S_0$	14.00
37 Rb	(K,L,M) $4s^2 4p^6 5s^1$	$^2S_{1/2}$	4.18

Element	Structure	Ground state	Ionization potential (eV)
38 Sr	(K,L,M) $4s^2\,4p^6\,5s^2$	$^1S_0$	5.69
39 Y	(K,L,M) $4s^2\,4p^6\,4d^1\,5s^2$	$^2D_{3/2}$	6.38
40 Zr	(K,L,M) $4s^2\,4p^6\,4d^2\,5s^2$	$^3F_2$	6.84
41 Nb	(K,L,M) $4s^2\,4p^6\,4d^4\,5s^1$	$^6D_{1/2}$	6.88
42 Mo	(K,L,M) $4s^2\,4p^6\,4d^5\,5s^1$	$^7S_3$	7.10
43 Tc	(K,L,M) $4s^2\,4p^6\,4d^5\,5s^2$	$^6S_{3/2}$	7.28
44 Ru	(K,L,M) $4s^2\,4p^6\,4d^7\,5s^1$	$^5F_5$	7.36
45 Rh	(K,L,M) $4s^2\,4p^6\,4d^8\,5s^1$	$^4F_{9/2}$	7.46
46 Pd	(K,L,M,$\overline{N}$)	$^1S_0$	8.33
47 Ag	(K,L,M,$\overline{N}$) $5s^1$	$^2S_{1/2}$	7.57
48 Cd	(K,L,M,$\overline{N}$) $5s^2$	$^1S_0$	8.99
49 In	(K,L,M,$\overline{N}$) $5s^2\,5p^1$	$^2P_{1/2}$	5.79
50 Sn	(K,L,M,$\overline{N}$) $5s^2\,5p^2$	$^3P_0$	7.34
51 Sb	(K,L,M,$\overline{N}$) $5s^2\,5p^3$	$^4S_{3/2}$	8.64
52 Te	(K,L,M,$\overline{N}$) $5s^2\,5p^4$	$^3P_2$	9.01
53 I	(K,L,M,$\overline{N}$) $5s^2\,5p^5$	$^2P_{3/2}$	10.45
54 Xe	(K,L,M,$\overline{N}$) $5s^2\,5p^6$	$^1S_0$	12.13
55 Cs	(K,L,M,$\overline{N}$) $5s^2\,5p^6\,6s^1$	$^2S_{1/2}$	3.89
56 Ba	(K,L,M,$\overline{N}$) $5s^2\,5p^6\,6s^2$	$^1S_0$	5.21
57 La	(K,L,M,$\overline{N}$) $5s^2\,5p^6\,5d^1\,6s^2$	—	5.61
58 Ce	(K,L,M,$\overline{N}$) $4f^2\,5s^2\,5p^6\,6s^2$	—	—
59 Pr	(K,L,M,$\overline{N}$) $4f^3\,5s^2\,5p^6\,6s^2$	—	—
60 Nd	(K,L,M,$\overline{N}$) $4f^4\,5s^2\,5p^6\,6s^2$	—	—
61 Pm	(K,L,M,$\overline{N}$) $4f^5\,5s^2\,5p^6\,6s^2$	~	—
62 Sm	(K,L,M,$\overline{N}$) $4f^6\,5s^2\,5p^6\,6s^2$	~	—
63 Eu	(K,L,M,$\overline{N}$) $4f^7\,5s^2\,5p^6\,6s^2$	~	—
64 Gd	(K,L,M,$\overline{N}$) $4f^7\,5s^2\,5p^6\,5d^1\,6s^2$	~	—
65 Tb	(K,L,M,$\overline{N}$) $4f^9\,5s^2\,5p^6\,6s^2$	—	—
66 Dy	(K,L,M,$\overline{N}$) $4f^{10}\,5s^2\,5p^6\,6s^2$	—	—
67 Ho	(K,L,M,$\overline{N}$) $4f^{11}\,5s^2\,5p^6\,6s^2$	—	—
68 Er	(K,L,M,$\overline{N}$) $4f^{12}\,5s^2\,5p^6\,6s^2$	—	—
69 Tm	(K,L,M,$\overline{N}$) $4f^{13}\,5s^2\,5p^6\,6s^2$	—	—
70 Yb	(K,L,M,$\overline{N}$) $4f^{14}\,5s^2\,5p^6\,6s^2$	—	—
71 Lu	(K,L,M,N) $5s^2\,5p^6\,5d^1\,6s^2$	$^2D_{3/2}$	—
72 Hf	(K,L,M,N) $5s^2\,5p^6\,5d^2\,6s^2$	$^3F_2$	7
73 Ta	(K,L,M,N) $5s^2\,5p^6\,5d^3\,6s^2$	$^4F_{3/2}$	7.88
74 W	(K,L,M,N) $5s^2\,5p^6\,5d^4\,6s^2$	$^5D_0$	7.98
75 Re	(K,L,M,N) $5s^2\,5p^6\,5d^5\,6s^2$	$^6S_{3/2}$	7.87
76 Os	(K,L,M,N) $5s^2\,5p^6\,5d^6\,6s^2$	$^5D_4$	8.7
77 Ir	(K,L,M,N) $5s^2\,5p^6\,5d^7\,6s^2$	$^4F_{9/2}$	9
78 Pt	(K,L,M,N) $5s^2\,5p^6\,5d^9\,6s^1$	$^3D_3$	9.0
79 Au	(K,L,M,N) $5s^2\,5p^6\,5d^{10}\,6s^1$	$^2S_{1/2}$	9.22
80 Hg	(K,L,M,N) $5s^2\,5p^6\,5d^{10}\,6s^2$	$^1S_0$	10.43
81 Tl	(K,L,M,N) $5s^2\,5p^6\,5d^{10}\,6s^2\,6p^1$	$^2P_{1/2}$	6.10
82 Pb	(K,L,M,N) $5s^2\,5p^6\,5d^{10}\,6s^2\,6p^2$	$^3P_0$	7.42

31

ELECTRON SHELL STRUCTURE

Element	Structure	Ground state	Ionization potential (eV)
83 Bi	$(K,L,M,N)\ 5s^2\,5p^6\,5d^{10}\,6s^2\,6p^3$	$^4S_{3/2}$	7.29
84 Po	$(K,L,M,N)\ 5s^2\,5p^6\,5d^{10}\,6s^2\,6p^4$	$^3P_2$	8.43
85 At	$(K,L,M,N)\ 5s^2\,5p^6\,5d^{10}\,6s^2\,6p^5$	—	—
86 Em	$(K,L,M,N)\ 5s^2\,5p^6\,5d^{10}\,6s^2\,6p^6$	$^1S_0$	10.75
87 Fr	$(K,L,M,N)\ 5s^2\,5p^6\,5d^{10}\,6s^2\,6p^6\,7s^1$	$^2S_{1/2}$	—
88 Ra	$(K,L,M,N)\ 5s^2\,5p^6\,5d^{10}\,6s^2\,6p^6\,7s^2$	$^1S_0$	5.23
89 Ac	$(K,L,M,N)\ 5s^2\,5p^6\,5d^{10}\,6s^2\,6p^6\,6d^1\,7s^2$	—	—
90 Th	$(K,L,M,N)\ 5s^2\,5p^6\,5d^{10}\,6s^2\,6p^6\,6d^2\,7s^2$	—	—
91 Pa	$(K,L,M,N)\ 5s^2\,5p^6\,5d^{10}\,5f^2\,6s^2\,6p^6\,6d^1\,7s^2$	—	—
92 U	$(K,L,M,N)\ 5s^2\,5p^6\,5d^{10}\,5f^3\,6s^2\,6p^6\,6d^1\,7s^2$	—	—

MOLECULAR BOND CONSTANTS

Bond	Bond length r_0 (nm)	Force constant $10^2\ \mathrm{N\,m^{-1}}$
H — H	0.0742	5.71
C — C	0.1573	4.50
C = C	0.1353	9.6
C ⁓ C	0.1395	—
C ≡ C	0.1207	15.7
N — N	0.1453	3.6
N = N	0.124	—
N ≡ N	0.1095	23.0
O — O	0.148	3.8
O = O	0.1208	11.8
F — F	0.1418	4.5
Cl — Cl	0.1988	3.3
H — F	0.0917	9.65
H — Cl	0.1275	5.16
H — Br	0.1414	4.10
C — H	0.109	4.88
C — N	0.148	4.9
C ≡ N	0.1157	17.7
C — O	0.1427	5.3
C = O	0.1225	12.1
C ≡ O	0.1128	19.0
N — H	0.1014	6.5
N — N	0.1453	3.6
N = O	0.137	—
O — H	0.0958	9.57

Bond length, $1\,\mathrm{nm} = 10\,\text{Å}$;

force constant, $1\,\mathrm{N\,m^{-1}} = 10^3\ \mathrm{dyne\,cm^{-1}}$

Force constant $= (\partial^2 E_p/\partial r^2)_{r_0}$

Linear molecules, bond angles 180°:

CO_2 CS_2 HCN C_2H_2 C_2N_2 N_2O

Symmetrical-top molecules:

NH_3 CH_3X C_2H_4 BF_3 N_3H C_2H_6 H_2CO

NH_3 ∢ HNH 106.8°		H_2CO ∢ HCH 123.4°
BF_3 ∢ FBF 120°		N_3H ∢ NNH 110.9°
CH_3X ∢ HCH (mean) 112° (X halide)		PH_3 ∢ HPH 99°
C_2H_4 ∢ HCH 119.9°		C_3H_4 ∢ CCC 180°
C_2H_6 ∢ HCH 112.2°		

Spherical-top molecules:

Bond angles 109.5° CH_4 CD_4 SiH_4 GeH_4

Bond angles 90° SF_6

Assymetric-top molecules:

H_2O ∢ HOH 104.5°	H_2S ∢ HSH 92.5°
SO_2 ∢ OSO 120°	CH_3OH ∢ COH 106°

TRANS-URANIUM ELEMENTS

Name and symbol		Nuclide	Half-life
Neptunium	Np	$^{237}_{93}$Np	2.2 10^6 y
Plutonium	Pu	$^{239}_{94}$Pu	2.43 10^4 y
Americium	Am	$^{241}_{95}$Am	473 y
Curium	Cm	$^{242}_{96}$Cm, $^{246}_{96}$Cm	162.5 d, 7 10^3 y
Berkelium	Bk	$^{247}_{97}$Bk	290 d
Californium	Cf	$^{249}_{98}$Cf	470 y
Einsteinium	Es	$^{254}_{99}$Es	1 y
Fermium	Fm	$^{250}_{100}$Fm	30 min
Mendelevium	Md	$^{256}_{101}$Md	30 min
(Nobelium	No ?)	$_{102}$No	3 s
Lawrencium	Lw	$_{103}$Lw	—

This list is not complete. The most abundant or the longest-lived nuclide only is given.

NATURALLY OCCURRING ISOTOPES

Given in order of abundance. ( ) indicates an abundance less than 1%. [ ] indicates a naturally occurring radioactive isotope.

Element		Isotopes (Mass numbers)	Element		Isotopes (Mass numbers)
1	H	1 (2)	44	Ru	102 101 104 100 99 96 98
2	He	4 (3)			
3	Li	7 6	45	Rh	103
4	Be	9	46	Pd	106 108 105 110 104 (102)
5	B	11 10			
6	C	12 13 [(14)]	47	Ag	107 109
7	N	14 (15)	48	Cd	114 112 111 110 113 116 106 (108)
8	O	16 (18) (17)			
9	F	19	49	In	[115] 113
10	Ne	20 22 (21)	50	Sn	120 118 116 119 117 124 122 112 (114) (115)
11	Na	23			
12	Mg	24 26 25	51	Sb	121 123
13	Al	27	52	Te	130 128 126 125 124 122 (123)
14	Si	28 29 30			
15	P	31	53	I	127
16	S	32 34 (33) (36)	54	Xe	132 129 131 134 136 130 128 (124) (126)
17	Cl	35 37			
18	A	40 (36) (38)	55	Cs	133
19	K	39 41 [(40)]	56	Ba	138 137 136 135 134 (130) (132)
20	Ca	40 44 (42) (48) (43) (46)			
			57	La	139 [(138)]
21	Sc	45	58	Ce	140 [142] 138 (136)
22	Ti	48 46 47 49 50	59	Pr	141
23	V	51 (50)	60	Nd	142 [144] 146 143 145 148 150
24	Cr	52 53 50 54			
25	Mn	55	62	Sm	152 154 [147] 149 148 150 144
26	Fe	56 54 57 (58)			
27	Co	59	63	Eu	153 151
28	Ni	58 60 62 61 64	64	Gd	158 160 156 157 155 154 (152)]
29	Cu	63 65			
30	Zn	64 66 68 67 (70)	65	Tb	159
31	Ga	69 71	66	Dy	164 162 163 161 160 (158) (156)
32	Ge	74 72 70 73 76			
33	As	75	67	Ho	165
34	Se	80 78 76 82 77 (74)	68	Er	166 168 167 170 164 (162)
35	Br	79 81	69	Tm	169
36	Kr	84 86 83 82 80 (78)	70	Yb	174 172 173 171 176 170 (168)
37	Rb	85 [87]	71	Lu	175 [176]
38	Sr	88 86 87 (84)	72	Hf	180 178 177 179 176 (174)
39	Y	89			
40	Zr	90 92 94 91 96	73	Ta	181 (180)
41	Nb	93	74	W	184 186 182 183 (180)
42	Mo	98 96 95 92 97 94 100	75	Re	[187] [185]

NATURALLY OCCURRING ISOTOPES

Element		Isotopes (Mass numbers)	Element		Isotopes (Mass numbers)
76	Os	192 190 189 188 187 186 (184)	81	Tl	205 203
			82	Pb	208 206 207 [204]
77	Ir	193 191	83	Bi	209
78	Pt	195 194 196 198 (192) [(190)]	90	Th	232
			92	U	[238] [(235)] [(234)]
79	Au	197			
80	Hg	202 200 199 201 198 204 (196)			

NUCLEAR CONSTANTS

1 atomic mass unit (u) $= 1.660\,53 \times 10^{-27}$ kg

$\qquad\qquad\qquad\quad = 931.481$ MeV

$1\,\text{MeV} = 1.602\,19 \times 10^{-13}$ J

Rest masses (u units):

Neutron	1.008 665 2	Proton	1.007 276 6
Deuteron	2.013 554	Electron	$5.485\,93 \times 10^{-4}$

Atomic masses (u units):

1	H	1.007 823	11	B	11.009 299
2	H	2.014 097	12	C	12.000 000
3	H	3.016 041	13	C	13.003 351
3	He	3.016 021	14	C	14.003 243
4	He	4.002 604	14	N	14.003 076
6	Li	6.015 114	15	N	15.000 108
7	Li	7.015 999	16	O	15.994 930
9	Be	9.012 186	17	O	16.999 144
10	B	10.012 940	18	O	17.999 168

Approximate radius of nucleus $= 1.45 \times 10^{-15} A^{1/3}$ m

$A =$ mass number = total number of nucleons $= N + Z$.

$N =$ number of neutrons; $Z =$ number of protons.

Empirical mass formula (u units):

$$\text{Mass} = 0.993\,64\,N + 0.992\,79\,Z + 0.0141\,(N + Z)^{2/3} +$$
$$+ 0.021\,(N - Z)^2/(N + Z) + 0.000\,63\,Z(Z - 1)/(N + Z)^{1/3}$$
$$- 0.073\,[(-1)^N + (-1)^Z]/(N + Z)$$

$\qquad =$ mass of atom with N neutrons, Z protons, Z orbital electrons.

ELASTIC CONSTANTS OF SOLIDS

The elastic constants of a sample depend to a great extent on its past history, crystalline structure, etc., so that the values below can only be taken as approximate.

	Bulk modulus (K) N m^{-2}	Young's modulus (E) N m^{-2}	Shear modulus (G) N m^{-2}	Poisson's Ratio σ
	10^{10}	10^{10}	10^{10}	
Metals				
Aluminium	7.5	7.0	2.5	0.34
Bismuth	3.0	3.2	1.2	0.33
Cadmium	4.2	5.0	2.1	0.30
Copper	13.5	11.0	4.4	0.34
Gold	16.5	8.0	2.8	0.42
Iron (wrought)	16.0	21.0	7.7	0.28
Iron (cast)	9.5	11.0	5.0	0.27
Lead	4.1	1.6	0.6	0.44
Magnesium	3.3	4.1	1.7	—
Nickel	17.0	21.0	7.8	0.30
Platinum	24.5	17.0	6.3	0.39
Silver	10.5	7.7	2.8	0.37
Steel (cast)	17.0	20.0	7.5	0.28
Steel (mild)	16.0	22.0	8.0	0.28
Tantalum	20.0	19.0	6.9	0.34
Tin	5.3	5.3	1.9	0.33
Tungsten	30.0	39.0	15.0	0.28
Zinc	3.5	8.0	3.6	0.23
Alloys				
Brass	6.0	9.0	3.5	0.35
Bronze	9.0	10.5	3.7	0.36
Bronze (phosphor)	—	12.0	4.4	0.38
German silver	15.0	11.0	4.5	0.37
Be copper	—	4.7	12.7	—
Miscellaneous				
Glass (crown)	5.0	6.0	2.5	0.25
Indiarubber	—	0.05	0.000 15	0.48
Quartz (fibre)	1.5	5.4	3.0	—
Wood (Oak)	—	1.3	—	—
Wood (Deal)	—	0.9	—	—
Wood (Teak)	—	1.7	—	—

For plastics and semiconductors see separate tables pages 38, 39 and 29.

COMPRESSIBILITY OF LIQUIDS

The quantity given is $1/V_0 \cdot \partial V/\partial P$, where V_0 is the volume at $0°C$ and 1 atmosphere pressure. The unit of pressure is $1\,\mathrm{kgf\,cm^{-2}}$. The value of V/V_0 is also given, in brackets. All values (except that for mercury) for $20°C$.

Substance	Pressure		
	1	500	1000
Acetone	120×10^{-6} (1.0279)	61×10^{-6} (0.9829)	51×10^{-6} (0.9553)
Alcohol– Amyl	89×10^{-6} (1.0181)	60×10^{-6} (0.9800)	45×10^{-6} (0.9526)
Ethyl (ethanol)	104×10^{-6} (1.0212)	62×10^{-6} (0.9794)	53×10^{-6} (0.9506)
Methyl (methanol)	113×10^{-6} (1.0283)	64×10^{-6} (0.9823)	53×10^{-6} (0.9530)
Propyl	91×10^{-6} (1.0173)	65×10^{-6} (0.9780)	46×10^{-6} (0.9498)
Carbon disulphide	91×10^{-6} (1.0235)	57×10^{-6} (0.9865)	47×10^{-6} (0.9586)
Ether	184×10^{-6} (1.0315)	83×10^{-6} (0.9681)	60×10^{-6} (0.9363)
Mercury (22°C)	3.95×10^{-6} (1.003 98)	3.89×10^{-6} (1.002 02)	3.83×10^{-6} (1.000 07)
Water	45.3×10^{-6} (1.0016)	38.1×10^{-6} (0.9808)	33.6×10^{-6} (0.9630)
$1\,\mathrm{kgf\,cm^{-2}} \;=\; 98\,067\,\mathrm{N\,m^{-2}} \;=\; 0.9672\,\mathrm{atm.}$			

	Units	Casein	Cellul-acetate	Cellul-aceto-butyrate	Cellul-nitrate	Methyl.methac
(a) Thermosetter (b) Thermoplastic	–	(b)	(b)	(b)	(b)	(b)
Density	10^3 kg m^{-3}	1.35	1.27–1.37	1.15–1.23	1.35–1.60	1.18
Hardness	Brinell no.	23	8–15 (10 kg)	6–12	8–11 (10 kg)	18–20
Youngs Modulus	10^9 N m^{-2}	3.5–3.9	1–3	0.4–2.5	1.4–2.8	2.8–4.1
Softening point	°C	95–130	65–125	60–120	70–90	60–70
Thermal expansion	10^{-6} °C^{-1}	65	140	120–160	100–160	80
Specific heat capacity	10^3 J kg^{-1} K^{-1}	1.5	1.2–1.8	1.5	1.5	1.5
Thermal conductivity	W m^{-1} K^{-1}	0.17	0.25–0.35	0.3	0.13–0.20	0.15–0.20
Refractive index	–	1.47–1.50	1.47–1.50	1.48	1.50	1.49
Volume resistivity	Ω m	–	10^8–10^{10}	10^8–10^{10}	10^8–10^9	10^{13}–10^{15}
Dielectric constant 50 s^{-1}	–	–				3.0–3.1
Dielectric constant 10^3 s^{-1}	–	–	3–6	3.6	7–8	–
Dielectric constant 10^6 s^{-1}	–	6.2–6.8			6.15	3.1–3.3
Power factor	–	0.052	0.2–0.6	0.02	0.07–0.15	0.06
Solvents	–	none	ketones esters	ketones esters	ketones esters	ketones esters arom.hyd
Machining qualities	–	good	good	good	good	excellent

38

Neoprene	Nylon	Phenol-formald	Poly-ethylene	Poly-styrene	Poly-tetra-fluoro ethylene	Poly-vinyl chloride	Poly-vinyl chloride acetate
(b)	(b)	(a)	(b)	(b)	(b)	(b)	(b)
1.24	1.14	1.3	0.95	1.05–1.07	2.2	1.3–1.4	1.34–1.37
–	23	30–45	1–2	20–30	low	2–50	12–25
–	3.1	5–7	low	3.2–3.5	low	low to 4.0	2.4–2.8
–	200–250	none	110	90–120	none	–	60–65
200	100	80	250	70–100	50	60–80	40–150
2.0	2.3	1.2–1.7	2.3	1.3	1.1	1.1	1.1
0.20	0.25	0.13–0.20	0.30–0.35	0.09	0.25	0.13–0.20	0.15–0.17
1.56	1.55	1.5–1.7	1.52	1.59–1.60	1.37	1.53	1.53
10^9-10^{11}	10^{11}	$10^{10}-10^{11}$	10^{15}	$10^{15}-10^{16}$	10^{17}	$10^{11}-10^{12}$	$10^{12}-10^{14}$
2.1	3.2	5–6	24	2.6	–	3.4–4.0	3.1–3.6
–	–	–	–	2.6	–	–	–
–	–	–	–	2.6	–	–	–
0.04	0.01–0.02	0.05–0.20	0.0005	0.0001	–	0.02–0.15	0.01–0.04
chlorin. aromatic hydroc.	none	none	none	chlorin. aromatic hydroc.	none	ketones esters	ketones esters
–	poor	fair to good	fair	poor to moderate	fair	fair to good	good to excellent

DENSITIES (ρ)

At ordinary room temperature, 17°–23°C

Substance	Density, 10^3 kg m⁻³	Substance	Density, 10^3 kg m⁻³
Aluminium	2.70	Rubidium	1.53
Antimony	6.62	Ruthenium	12.3
Arsenic (metallic)	5.73	Samarium	7.75
Barium	3.5	Selenium (amorph)	4.8
Beryllium	1.84	Silicon (amorph.)	2.35
Bismuth	9.78	Silicon (cryst.)	2.42
Boron	2.33	Silver	10.5
Cadmium	8.65	Sodium	0.97
Cæsium	1.87	Strontium	2.56
Calcium	1.54	Sulphur (amorph.)	1.92
Carbon (graphite)	2.22	Tantalum	16.6
(diamond)	3.514	Tellurium (cryst.)	6.25
Cerium	6.80	Thallium	11.86
Chromium	7.10	Thorium	11.3
Cobalt	8.7	Tin	7.3
Copper	8.89	Titanium	4.5
Gallium	5.93	Tungsten (wire)	19.3
Germanium	5.46	Uranium	19.05
Gold	19.3	Vanadium	6.0
Indium	7.28	Zinc	7.1
Iodine	4.94	Zirconium	6.4
Iridium	22.42	*Alloys*	
Iron (pure)	7.88		
(wrought)	7.85	Bell metal	8.7
(cast)	7.6	Brass	8.4—8.7
(steel)	7.7	Bronze	8.8—8.9
Lanthanum	6.15	(phosphor)	8.8
Lead	11.34	Constantan	8.88
Lithium	0.534	Be-copper	8.24
Magnesium	1.74	Invar	8.00
Manganese	7.41	Magnalium	2.0—2.5
Mercury	14.19	Manganin	8.50
(solid, −39°C)		Steel	7.8
Molybdenum	10.1	Wood's metal	9.5—10.5
Neodymium	6.96	*Miscellaneous*	
Nickel	8.8		
Osmium	22.5	Asbestos	2.0—2.8
Palladium	12.2	Cork	0.22—0.26
Phosphorus (red)	2.20	Ebonite	1.15
(yellow)	1.83	Glass	2.4—2.8
Platinum	21.45	Ice, 0°C	0.917
Potassium	0.86	Mica	2.6—3.2
Praseodymium	6.48	Perspex	1.19
Rhodium	12.44	Polystyrene	1.06

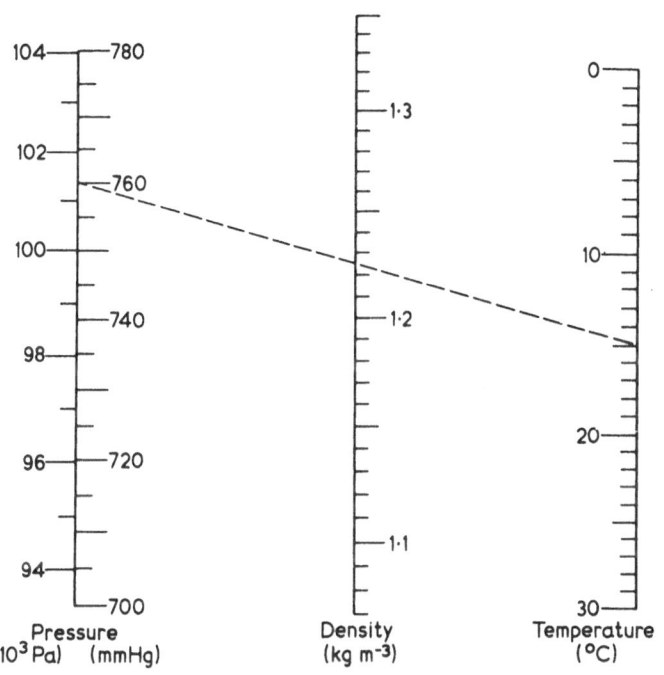

Nomogram 5. Density of dry air.

A straight edge placed across the diagram passes through three inter-related quantities. Thus at 15°C and at a pressure of 760 mmHg the density of dry air is 1.225 kg m⁻³.

41

DENSITIES (ρ)

Substance	Density, 10^3 kg m^{-3}	Substance	Density, 10^3 kg m^{-3}
Miscellaneous (cont,)		*Liquid Gases*	
Fused silica	2.1—2.2	*(Standard atm. pressure)*	
Paraffin wax	0.9	Hydrogen 20.39 K	0.0710
Woods		Helium-3 2.178 K	0.147
Ash	0.6—0.8	Helium-4 4.20 K	0.125
Balsa	0.12—0.20	Nitrogen 77.31 K	0.808
Beech	0.7—0.9	Oxygen 90.18 K	1.145
Elm	0.5—0.6	Air 79 K	21% O$_2$ 0.920
Mahogany	0.6—0.8		94% O$_2$ 1.133
Oak	0.6—0.9		
Teak	0.7—0.9	*Gases*	kg m^{-3}
			at s.t.p.
Liquids, 15°C		Air	1.2928
Acetone	0.792	Ammonia	0.7714
Ethanol	0.791	Argon	1.784
Methanol	0.810	Bromine	7.139
Aniline	1.022	Carbon monoxide	1.2500
Benzene	0.899	Carbon dioxide	1.9770
Ether	0.736	Chlorine	3.214
Glycerine	1.26	Fluorine	1.69
Oil		Helium	0.1785
lubricating	0.90—0.92	Hydrochloric acid	1.639
olive	0.92	Hydrogen	0.0899
paraffin	0.8	Hydrogen sulphide	1.539
Seawater	1.025	Krypton	3.74
Heavy water	1.1058	Methane	0.7167
Turpentine	0.87	Neon	0.900
Normal ½H$_2$SO$_4$	1.0304	Nitrogen	1.251
Normal HCl	1.0162	Oxygen	1.429
Normal HNO$_3$	1.0322	Xenon	5.90
Normal NaOH	1.0414		
Normal NaCl	1.0388		
Normal KOH	1.048		
Normal KCl	1.0446		

Temperature °C	Water 10^3 kg m^{-3}	Mercury, 10^3 kg m^{-3}	Temperature °C	Water, 10^3 kg m^{-3}	Mercury, 10^3 kg m^{-3}
0	0.999 841	13.5951	50	0.988 04	13.4726
1 or 7	0.999 902	—	60	0.983 21	13.4483
2 or 6	0.999 941	—	70	0.977 79	13.4240
3 or 5	0.999 965	—	80	0.971 80	13.3998
4	0.999 972	—	90	0.965 31	13.3756
10	0.999 700	13.5705	100	0.958 35	13.3515
20	0.998 203	13.5459	150	0.917 3	13.2315
30	0.995 646	13.5214	200	0.862 8	13.1120
40	0.992 21	13.4970			

VISCOSITIES

The unit of dynamic viscosity (η) is the pascal second (Pa s). Values throughout the tables are in 10^{-6} Pa s, unless otherwise indicated. (The CGS unit is the poise (P) = 10^{-1} Pa s). The ratio dynamic viscosity/density (η/ρ) is called kinematic viscosity (ν); the SI unit is $1\,m^2\,s^{-1}$ and has so far received no name; the CGS unit is the stokes (S) = $10^4\,m^2\,s^{-1}$.

Water.

Temperature °C	0	10	20	30	40	50	60	70	80	90	100
Viscosity	1787	1304	1002	798	654	548	467	405	355	316	283

Mercury.

Temperature °C	0	50	100
Viscosity	1685	1409	1226

Miscellaneous liquids.

Substance	Temp. °C	Viscosity, 10^{-6} Pa s	Substance	Temp. °C	Viscosity, 10^{-6} Pa s
Acetone	20	327	Oils		
Alcohol			castor	20	986×10^3
ethyl	20	1196	linseed	30	331×10^2
methyl	20	592	olive	20	84×10^3
Benzene	20	648	Sulphuric acid	20	22×10^3
Carbon			Turpentine	20	1490
disulphide	20	370	Xylol		
Ether	20	234	ortho	20	807
Glycerine	20	149.5×10^4	meta	20	615
Nitric acid	10	1770	para	20	643

Mineral oil and silicone fluid.

Substance	Kinematic viscosity (ν) m²s⁻¹	
	Temp. 40°C	Temp. 100°C
Mineral oil		
S.A.E. 10	3 500	540
S.A.E. 20	7 400	850
S.A.E. 30	11 900	1130
S.A.E. 40	18 400	1480
S.A.E. 50	26 000	1900
Silicone fluid		
DC 200−350	$35 \cdot 10^3$	$14 \cdot 10^3$
DC 200−20	1500	700

Gases.

Substance	Temp. °C	Viscosity, 10^{-6} Pa s	Substance	Temp. °C	Viscosity, 10^{-6} Pa s
Air	20	18.1	Krypton		23.3
Ammonia	20	10.8	Methane	20	11.0
Argon	0	21.2	Neon	0	29.8
Carbon monox.	20	18.0	Nitric oxide	20	18.6
Carbon diox.	20	16.2	Nitrogen	20	17.6
Chlorine	20	13.5	Oxygen	20	20.5
Helium	0	19.2	Sulphur dioxide	20	13.2
Hydrogen	20	9.0	Water vapour	15	9.8
Hydrogen sulph.	20	13.0	Xenon	0	21.1
Hydrogen chlor.	0	14.0			

Viscosity of aqueous sucrose solutions

All values in 10^{-3} Pa s units

$t°C$ \ $W\%$	0	20	30	35	40	45	50	55	60	65	70
10	1.304	2.647	4.494	6.253	9.211	14.56	25.17	48.88	110.6	307.5	1159
11	1.267	2.562	4.328	6.002	8.809	13.86	23.85	46.05	103.6	286.5	1074
12	1.233	2.482	4.173	5.770	8.439	13.22	22.64	43.47	97.19	266.8	992.9
13	1.199	2.404	4.026	5.552	8.092	12.63	21.51	41.06	91.16	248.1	913.9
14	1.167	2.331	3.888	5.347	7.768	12.07	20.47	38.83	85.55	230.6	838.5
15	1.137	2.261	3.758	5.155	7.467	11.56	19.50	36.77	80.33	214.1	766.7
16	1.108	2.194	3.635	4.975	7.186	11.08	18.61	34.85	75.47	198.7	698.9
17	1.080	2.130	3.519	4.807	6.923	10.64	17.78	33.08	70.95	184.3	635.1
18	1.053	2.069	3.409	4.648	6.677	10.23	17.00	31.42	66.71	170.7	575.0
19	1.027	2.011	3.306	4.499	6.448	9.845	16.29	29.89	62.79	158.1	519.5
20	1.002	1.955	3.207	4.359	6.232	9.487	15.62	28.45	59.10	146.3	467.6

$W\%$ = percentage by weight of sucrose.

Viscosity of glycerol–water mixtures

All values in 10^{-3} Pa s units.

% weight glycerol	20°C	22°C	24°C	26°C	28°C	30°C
0	1.002	0.953	0.913	0.873	0.830	0.798
5	1.140	1.084	1.032	0.983	0.940	0.900
10	1.306	1.241	1.180	1.124	1.071	1.022
15	1.513	1.434	1 360	1.292	1.229	1.172
20	1.764	1.669	1.580	1.498	1.424	1.356
25	2.089	1.971	1.862	1.762	1.669	1.585
30	2.496	2.352	2.217	2.092	1.887	1.872
35	3.033	2.850	2.680	2.521	2.376	2.243
40	3.739	3.499	3.279	3.076	2.891	2.724
45	4.701	4.385	4.094	3.826	3.585	3.369
50	6.044	5.613	5.219	4.860	4.536	4.247
55	7.978	7.378	6.828	6.329	5.880	5.482
60	10.93	10.05	9.252	8.527	7.878	7.303
65	15.45	14.10	12.88	11.78	10.83	9.990
70	22.90	20.74	18.79	17.07	15.56	14.27
75	36.33	32.60	29.27	26.33	23.78	21.61
80	61.82	54.84	48.65	43.26	38.64	34.82
85	113.0	99.04	86.79	76.19	67.26	59.98
90	234.5	202.9	175.3	151.5	131.7	115.9
95	544.7	463.7	393.7	334.5	286.2	248.8
96	659.3	559.3	473.0	400.3	341.2	295.7
97	798.4	675.5	569.5	480.5	408.3	353.0
98	971.6	820.0	688.0	577.0	489.0	422.0
99	1195	1006	842.0	706.0	595.0	510.0
100	1495	1248	1039	864	725	622

SURFACE TENSION (γ)

Measured in $10^{-3}\,\mathrm{N\,m^{-1}}$ at 20°C In all cases the liquid is assumed to be in contact with air.

Substance	Surface tension	Substance	Surface tension
Acetone	23.7	Water 60°C	66.2
Alcohol		80	62.6
ethyl	22.3	100	58.8
methyl	22.6		
Amyl acetate	24.7	*Solutions*	
Aniline	43	Conc. H_2SO_4	55
Benzene	28.9	HNO_3	41
Chloroform	27.2		
Glycerine	64	For every 1 g of anhydrous salt
Mercury	(vac.) 475	per 100 g of water, add to the
	(in air, decrease	value for pure water approxi-
	with age) 500—400	mately
Olive oil	33		
Paraffin oil	26		
Turpentine	27		
Water 5°C	74.92	$CaCl_2$	0.29
10	74.22	$CuSO_4$	0.11
15	73.49	KCl	0.19
20	72.75	KOH	0.32
25	71.97	NaCl	0.28
30	71.18	NaOH	0.50
40	69.6	NH_4Cl	0.26

SOLUBILITIES

Solubility of Some Common Gases in Water

The value given is the volume of gas, at s.t.p., which is absorbed by 1 cm³ of water at 20°C under a pressure of 1 atmosphere.

Gas	Volume absorbed, cm³	Gas	Volume absorbed, cm³
Air	0.019	Hydrogen	0.0181
Ammonia	700	Hydrogen sulphide	2.6
Argon	0.038	Hydrochloric acid	440
Carbon dioxide	0.85	Neon	0.017
Carbon monoxide	0.023	Nitrogen	0.0154
Chlorine	2.3	Oxygen	0.031
Helium	0.0085	Sulphur dioxide	39

Solubility of Some Common Chemical Compounds in Water

The solubilities are expressed in g of anhydrous substance per 100 g of water.

Substance	Formula	Solubility at	
		15°C	100°C
Ammonium			
chloride	NH_4Cl	35.3	77.3
nitrate	NH_4NO_3	173	871
sulphate	$(NH_4)_2SO_4$	74.2	103.3
Boric acid	H_3BO_3	4.3	40
Copper sulphate	$CuSO_45H_2O$	19.0	75.4
Ferrous sulphate	$FeSO_47H_2O$	23.5	37.3 (90°)
Lead nitrate	$Pb(NO_3)_2$	52.4	138.8
Potash alum	$Al_2(SO_4)_3K_2SO_424H_2O$	5.0	109 (90°)
Potassium			
chloride	KCl	32.5	56.7
bichromate	$K_2Cr_2O_7$	9	95
hydroxide	KOH	107	178
iodide	KI	140	208
nitrate	KNO_3	26.3	246
permanganate	$KMnO_4$	5.4	22.2 (60°)
Silver nitrate	$AgNO_3$	196	952
Sodium			
carbonate	$Na_2CO_310H_2O$	17	45
bicarbonate	$NaHCO_3$	8.9	23.5
chloride	$NaCl$	35.9	39.8
hydroxide	$NaOH$	105	340
sulphate	$Na_2SO_410H_2O$	13.1	42
thiosulphate	$Na_2S_2O_35H_2O$	65.5	266
Zinc chloride	$ZnCl_2$	344	614
Zinc sulphate	$ZnSO_47H_2O$	50.7	78.5
Cane sugar	$C_{12}H_{22}O_{11}$	197	487

I.S.O. metric (mm).

	O/D	Core	T/cm	O/D	Core	T/cm	
	2.0	1.51	25.0	7.0	5.77	10.0	
	2.5	1.95	22.2	8.0	6.47	8.0	
	3.0	2.39	20.0	10.0	8.16	6.7	
	3.5	2.76	16.7	12.0	9.85	5.7	
	4.0	3.14	14.3	14.0	11.55	5.0	
	4.5	3.58	13.3	16.0	13.55	5.0	
	5.0	4.02	12.5	18.0	14.93	4.0	
	6.0	4.77	10.0	20.0	16.93	4.0	

B.A. (mm).

No.	O/D	Core	T/cm	No.	O/D	Core	T/cm
0	6.00	4.80	10.0	6	2.80	2.16	18.8
1	5.30	4.22	11.1	7	2.50	1.93	20.8
2	4.70	3.73	12.3	8	2.20	1.69	23.3
3	4.10	3.23	13.7	9	1.90	1.43	25.6
4	3.60	2.81	15.1	10	1.70	1.28	28.5
5	3.20	2.50	17.0				

Whitworth (inches).

Size (O/D)	Core	T/in	Size (O/D)	Core	T/in
1/8	0.093	40	1/2	0.393	12
3/16	0.134	24	9/16	0.456	12
1/4	0.186	20	5/8	0.509	11
5/16	0.241	18	3/4	0.622	10
3/8	0.295	16	7/8	0.733	9
7/16	0.346	14	1	0.840	8

U.N.F. (inches).

Size (O/D)	Core	T/in	Size (O/D)	Core	T/in
1/4	0.206	28	9/16	0.494	18
5/16	0.261	24	5/8	0.557	18
3/8	0.324	24	3/4	0.673	16
7/16	0.376	20	7/8	0.787	14
1/2	0.439	20	1	0.898	12

SCREWTHREAD DIMENSIONS

B.S.F. (inches).

Size (O/D)	Core	T/in	Size (O/D)	Core	T/in
3/16	0.148	32	9/16	0.483	16
1/4	0.201	26	5/8	0.534	14
5/16	0.254	22	3/4	0.643	12
3/8	0.311	20	7/8	0.759	11
7/16	0.366	18	1	0.872	10
1/2	0.420	16			

O/D = overall diameter   T/cm = threads per cm
T/in = threads per inch

Woodscrews

No.	Shank diameter (mm)	No.	Shank diameter (mm)
0	1.52	9	4.52
1	1.78	10	4.88
2	2.08	11	5.23
3	2.39	12	5.59
4	2.74	14	6.30
5	3.10	16	7.01
6	3.45	18	7.72
7	3.81	20	8.43
8	4.17		

Diameter of head 0.794 (No. + 2) mm

HEAT

measured in Pascal (Pa)

Temperature °C	Water	Mercury
Liquid air	—	3×10^{-25}
Solid CO_2	0.08	4×10^{-7}
– 20	(ice) 1.04×10^2	—
– 10	(ice) 2.60×10^2	—
0	6.11×10^2	0.025
10	12.3×10^2	0.067
20	23.3×10^2	0.16
30	42.4×10^2	0.37
40	73.7×10^2	0.81
50	123×10^2	1.7
60	199×10^2	3.3
70	312×10^2	6.4
80	473×10^2	12
90	701×10^2	21
100	1.013×10^5	36
150	4.76×10^5	3.7×10^2
200	15.6×10^5	23×10^2
300	86.1×10^5	329×10^2

$1\,Pa = 1\,N\,m^{-2} = 7.5\ 10^{-3}\,mmHg;\quad 1\,mmHg = 133.3\,Pa$

Substance	Vapour pressure (10^2 Pa) Temperature 20°C
Acetic acid	15.5
Acetone	247
Alcohol	
iso-amyl	3.1
ethyl (ethanol)	58.5
methyl (methanol)	118
Benzene	99.4
Bromine	229
Carbon disulphide	397
Carbon dioxide	57.2×10^3
Carbon tetrachloride	121
Chloroform	215
Ethyl acetate	97
Ethyl ether	589
Hydrogen sulphide	18×10^3
Sulphur dioxide	3.27×10^3

WET AND DRY BULB HYGROMETER TABLES

(Ventilated type hygrometer)

The tabulated values are relative humidities (%).

Depression of wet bulb °C	Dry bulb temperature °C								
	0	5	10	15	20	25	30	35	40
1	81	87	88	89	90	92	93	93	94
2	64	72	76	80	82	85	86	87	88
3	46	59	66	71	74	77	79	81	82
4	29	45	55	62	66	70	73	75	76
5	13	33	44	53	59	63	67	70	72
6	–	21	34	44	52	57	61	64	66
7	–	9	25	36	45	50	55	59	61
8	–	–	15	28	38	44	50	54	56
9	–	–	6	20	30	38	44	50	52
10	–	–	–	13	24	33	39	44	48

CRITICAL CONSTANTS OF GASES

Gas	Critical pressure (10^6 Pa)	Critical temperature (K)	Critical density (kg m^{-3})
Ammonia	11.34	405.7	236
Argon	4.86	150.74	531
Bromine	10.3	584	1184
Carbon dioxide	7.37	304.1	465
Carbon monoxide	3.5	132.92	301
Chlorine	7.71	417	567
Helium	0.2291	5.206	68
Hydrogen	1.295	33.20	31
Krypton	5.496	209.4	909
Methane	4.63	190.8	162
Neon	2.725	44.4	484
Nitrogen	3.393	126.0	311
Oxygen	5.06	154.6	420
Water	22.10	647.3	317
Xenon	5.89	289.7	1105

THERMAL CONSTANTS OF THE ELEMENTS

Melting points and boiling points at standard pressure, specific heat capacities, latent heat of fusion, expansion coefficient and thermal conductivity.

Substance	m.p. °C	b.p. °C	Sp. ht. cap. J kg⁻¹K⁻¹	Temp. range °C	Latent heat fusion J kg⁻¹ 10⁴	Exp. coeff. °C⁻¹ 10⁻⁶	Thermal cond. W m⁻¹ K⁻¹
Aluminium	660.1	2400	908	17–100	40	25	242
Antimony	630.5	1750	210	20	16.5	11	18.5
Argon	−189.4	−185.9	—	—	—	—	—
Arsenic	sub.	615	335	0–100	—	6	—
Barium	710	1600	290	18	—	—	—
Beryllium	1280	2700	1780	0–100	—	12	200
Bismuth	271	1560	112	17–100	5.5	13.5	8.7
Boron	2100	3700	1284	0–100	—	—	—
Bromine	−7.3	58	448	13–45	6.7	—	—
Cadmium	321	770	230	20	5.5	30	96
Caesium	28.5	680	200	1–25	1.7	97	—
Calcium	850	1450	623	0–20	—	—	—
Carbon	sub.	3550	690	11	—	5	—
Cerium	800	3468	188	0–100	—	—	—
Chlorine	−102	−34	946	0–24	—	—	—
Chromium	1900	2400	460	17–100	—	7.4	87
Cobalt	1492	2900	435	15–100	24	12	93
Copper	1083	2580	385	15–100	20	16.7	383
Fluorine	−220	−188	—	—	—	—	—
Gallium	29.8	2403	377	12–100	8	19.2	34
Germanium	937	2800	324	0–100	—	5.7	59

THERMAL CONSTANTS OF THE ELEMENTS

Substance	m.p. °C	b.p. °C	Sp. ht. cap. J kg⁻¹K⁻¹	Temp. range °C	Latent heat fusion J kg⁻¹10⁴	Exp. coeff. °C⁻¹ 10⁻⁶	Thermal cond. W m⁻¹K⁻¹
Gold	1063	2650	128	17–100	6.7	14	300
Helium	−272	−268.9	–	–	–	–	–
Hydrogen	−259	−252.7	–	–	–	–	–
Indium	156.6	2100	238	0–100	–	48	84
Iodine	113.5	184.4	226	9–98	5.0	90	–
Iridium	2443	4600	135	18–100	–	6.5	59
Iron	1535	2750	460	18–100	21	12	71
Krypton	−157.3	−153.4	–	–	–	–	–
Lanthanum	920	3470	188	0–100	–	–	–
Lead	327.3	1750	127	20–100	2.5	29	36
Lithium	183	1360	4573	0–100	–	60	71
Magnesium	651	1120	1030	17–100	30	26	154
Manganese	1260	2150	508	0–100	–	21	–
Mercury	−38.87	356.58	139.3	20	1.2	–	9
Molybdenum	2620	4700	301	15–93	–	5	142
Neodymium	1024	3100	188	0–100	–	–	–
Neon	−248.7	−246.0	–	–	–	–	–
Nickel	1453	2900	456	15–100	29	13	59
Niobium	2420	5000	–	–	–	–	53
Nitrogen	−210.0	−195.8	–	–	–	–	–
Osmium	2700	4600	130	19–98	–	7	87
Oxygen	−218.8	−182.97	–	–	–	–	–
Palladium	1552	3560	247	18–100	15	12	74
Phosphorus	44.2	282	790	7–30	2.1	124	–
Platinum	1769	3900	135	15–100	11.5	9.0	71
Potassium	63.2	760	790	0–56	6.3	80	97

Praseodymium	932	3000	192	0–100	—	—	—
Radium	700	1140	—	—	—	—	—
Rhodium	1960	3700	243	10–100	—	8.5	89
Rubidium	38.8	700	335	0	—	90	60
Ruthenium	2300	4150	255	0–100	—	10	105
Samarium	1050	1600	—	—	—	—	—
Selenium	217	688	324	22–62	35	26	0.24
Silicon	1410	2355	706	20	—	2.5	175
Silver	960.8	2190	234	15–100	10.5	19	414
Sodium	97.8	883	1184	0	11.3	70	135
Strontium	770	1400	—	—	—	—	—
Sulphur (rh)	112.8	444.6	682	15–45	3.8	90	0.25
(mono)	119	—	757	0–50	—	—	—
Tantalum	3000	5450	151	58	—	—	—
Tellurium	452	1390	201	15–100	—	6.5	56
Thallium	303	1460	138	20–100	2.9	17	50
Thorium	1842	4200	117	0–100	—	30	48
Tin	231.91	2350	225	20	5.8	12	40
Titanium	1670	3300	473	0–100	—	23	63
Tungsten	3380	5700	142	20–100	—	9	23
Uranium	1133	3800	117	0–100	—	4.3	185
Vanadium	1920	3400	481	0–100	—	—	28
Xenon	-112	-108	—	—	—	—	29
Yttrium	1490	3000	—	—	—	—	—
Zinc	419.505	908	387	20	10.5	29	111
Zirconium	1850	4400	280	0–100	—	—	22

THERMAL CONSTANTS
OF COMMON SUBSTANCES

Melting points

	m.p.		m.p.
Beeswax	62°C	Gun-metal	1010°C
Naphthalene	80	Invar	1500
Paraffin wax	50—60	Magnalium	610
Quartz (fused)	1670	Solder (hard)	abt. 900
Potassium nitrate	335	(soft)	abt. 180
Sodium chloride	801	Steel	1400
Brass	900	Wood's metal	65
Constantan	1290		

Boiling points

	Atm. press. b.p.		Atm. press. b.p.
Acetone	56.7°C	Carbon disulphide	46.2°C
Aniline	184.2	Carbon tetrachloride	76.7
Alcohol		Chloroform	61.2
ethyl (ethanol)	78.3	Ether	34.6
methyl (methanol)	64.7	Glycerol	290
Benzene	80.2	Turpentine	161

Melting points and boiling points of gaseous compounds

	m.p.	b.p.		m.p.	b.p.
Ammonia	−78°C	−33.5°C	Nitric oxide	−163.7°C	−151.8°C
Carbon dioxide	−57	−78.5	Nitrogen peroxide	−11	21.2
Carbon monoxide	−205	−191.5	Sulphur dioxide	−75.5	−10
Hydrogen chloride	−114	−85	Sulphuretted hydrogen	−85.5	−60.5
Methane	−183	−161.5			

Expansion coefficients

Solids	$°C^{-1} \times 10^{-6}$	Solids	$°C^{-1} \times 10^{-6}$
Brass	19	Glass Pyrex	3
Brick	9.5	Gun-metal	18.1
Celluloid	110	Invar	1
Concrete	12	Magnalium	24
Constantan	16	Nichrome	12
Be-copper	17	Quartz (fused)	0.42
Duralumin	22.6	Solder (soft)	25
German silver	18.4	Steel (mild)	12
Glass (soft)	8.5		

Expansion coefficients (cont.)

Liquids (Volume coeff.)

Ethanol	0.001 10
Methanol	0.001 22
Aniline	0.000 85
Benzene	0.001 24
Carbon disulphide	0.001 21
Carbon tetrachloride	0.001 23
Chloroform	0.001 27
Ether	0.001 63
Glycerol	0.000 53
Mercury	0.000 182
Hydrochloric acid	0.000 57

Liquids (Volume coeff.)

Oil	
olive	0.000 72
paraffin	0.000 90
turpentine	0.000 97
Sulphuric acid	0.000 57
Water 10°–20°C	0.000 15
20°–30°	0.000 25
30°–40°	0.000 35
40°–60°	0.000 46
60°–80°	0.000 59
80°–100°	0.000 70

Specific heat capacity

Solids — J kg^{-1}K^{-1}

Be-copper	420
Brass	389
Building stone	750–960
Constantan	420
Glass (soft)	670
(pyrex)	780
Gun metal	360
Ice (0°C)	2090
Invar	500
Magnalium	920
Marble	880
Paraffin wax	2900
Quartz (fused)	728
Rubber	1670
Solder (soft)	180
Steel	450
Wood's metal	146
Wood	1675

Liquids

Ethanol	2430
Methanol	2510
Aniline	2150
Benzene	1700
Brine (25 wt. % NaCl)	3210
Carbon disulphide	837

Liquids (cont.) — J kg^{-1} K^{-1}

Carbon tetrachloride	1000
Chloroform	980
Ether	2340
Glycerol	2430
Oil	
Paraffin	abt. 2180
Castor	2000
Olive	1970
Turpentine	1760

Gases and vapours

	C_p	C_p/C_v
Acetylene	1645	1.25
Air	1008	1.40
Ammonia	2190	1.31
Argon	527	1.66
Carbon monoxide	1050	1.40
Carbon dioxide	840	1.30
Chlorine	500	1.36
Helium	5230	1.64
Hydrogen	14270	1.41
Methane	2210	1.31
Nitrogen	1042	1.40
Oxygen	916	1.40
Sulphur dioxide	640	1.29
Sulphuretted hydrogen	1070	1.33

Specific heat capacity of water ($J\,kg^{-1}K^{-1}$)

0°C 4217	15°C 4186	30°C 4178	60°C 4184	90°C 4205
5°C 4202	20°C 4182	40°C 4178	70°C 4189	99°C 4215
10°C 4192	25°C 4179	50°C 4180	80°C 4196	

Latent heats

Fusion	$10^2\,J\,kg^{-1}\,K^{-1}$	*Vaporization*	$10^2\,J\,kg^{-1}K^{-1}$
Acetic acid	1860	Ethanol	8540
Beeswax	1770	Methanol	11090
Benzene	1268	Benzene	3926
Bismuth	515	Carbon disulphide	3530
Glycerol	1880	Carbon tetrachloride	1930
Ice	3335	Ether	3750
Lead	247	Mercury	2760
Sulphur	397	Nitrogen	2000
Tin	582	Oxygen	2135
		Water	22564

Thermal Conductivities

All values have as the unit $10^{-2}\,W\,m^{-1}\,K^{-1}$. ($1\,W\,m^{-1}\,K^{-1} = 0.002\,39\ cal\,cm^{-1}\,s^{-1}\,°C^{-1}$).

Solids

	Bulk density $10^3\,kg\,m^{-3}$	Cond.		Bulk density $10^3\,kg\,m^{-3}$	Cond.
Asbestos paper, wool	0.5	15	Sheepswool	0.08	4
Bitumen	1.3	17	Silk fabric	—	4
Brass	8.6	10 900	Slag wool	0.20	4.6
Brick	1.6	50	Slate	3.0	138
Concrete	2.2	abt. 105	Snow	0.25	15
Constantan	8.9	2 260	Soda glass	2.6	71
Be-copper	8.24	2 550	Soil (dry)	—	18
Cork	0.20	5	Steel	7.8	4 600
Cotton wool	0.08	4.2	Steel wool	0.10	8
Ebonite	1.20	16	Wood		
Felt (hair)	0.27	3.8	balsa	0.14	5.5
Glass wool	0.22	3.8	common, perp. to grain	—	11—16
Ice	0.92	220	common, par. to grain	—	22—35
Kapok	0.015	3.3	sawdust	0.20	6
Paper	—	5			
Polystyrene (expanded)	0.05	3.3			
Pyrex	2.2	110			
Quartz (fused)	2.2	100			
Rubber (pure)	0.92	13			

Thermal conductivities (cont.)

Liquids (20°C)		Gases (0°C)	
Ethanol	17.6	Air	2.39
Methanol	21.0	Carbon dioxide	1.38
Aniline	17.2	Hydrogen	15.7
Benzene	13.8	Methane	3.10
Paraffin oil	15.1	Nitrogen	2.39
Turpentine	13.8	Oxygen	2.39
Water	60		

FIXED POINTS OF THE TEMPERATURE SCALE

The pressure $p_0 = 1$ standard atmosphere $= 101\,325$ N m^{-2}.

The boiling point of oxygen $-182.970°C$

for a pressure p, $t_p = -182.970 + 9.530(p/p_0 - 1)$
$$- 3.72(p/p_0 - 1)^2 + 2.2(p/p_0 - 1)^3.$$

The triple point of water $= 0.01°C$.

The boiling point of water $100°C$

for a pressure p, $t_p = 100 + 28.012(p/p_0 - 1)$
$$- 11.64(p/p_0 - 1)^2 + 7.1(p/p_0 - 1)^3.$$

The boiling point of sulphur $444.600°C$

for a pressure p, $t_p = 444.600 + 69.01(p/p_0 - 1)$
$$- 27.48(p/p_0 - 1)^2 + 19.14(p/p_0 - 1)^3.$$

The melting point of silver $960.8°C$.

The melting point of gold $1063°C$

The zero point of the Celsius scale is 273.15 K.

The triple point of water on the absolute thermodynamic scale is 273.16 K.

RADIATION CONSTANTS

Black-body

The Stefan–Boltzmann Constant (σ)

The total radiant energy emitted by a black body per unit area per unit time $= \sigma T^4$,
where

$$\sigma = 5.6696 \times 10^{-8} \text{ W m}^{-2} \text{ K}^{-4},$$
$$T = \text{absolute temperature of black body.}$$

RADIATION CONSTANTS

Wien's Law

$$T \lambda_{max} = \text{a constant} = 2.8980 \ 10^{-3} \text{ m K,}$$

where

T = absolute temperature of black body,

λ_{max} = wavelength of the most copiously emitted radiant energy.

Planck's Law

Expresses the distribution of energy over the wavelengths of radiation emitted by a black body:

$$E_\lambda = c_1 \lambda^{-5} (e^{-c_2/\lambda T} - 1)^{-1}$$

where $c_2 = hc/k = 1.438\,83 \times 10^{-2}$ m K.

For the amount of linearly polarized radiant energy emitted by a black body per unit area per unit solid angle per unit range of wavelength per unit time:

$$c_1 = hc^2 = 5.955\,34 \times 10^{-17} \text{ W m}^2 \text{ sr}^{-1}.$$

For the energy density, per unit range of wavelength, of radiation contained in an enclosed cavity when in temperature equilibrium with the walls

$$c_1 = 8\pi ch = 4.992\,58 \times 10^{-24} \text{ J m.}$$

λ = wavelength in m, T = absolute temperature of black body.

Tungsten

Temperature K	Colour temperature K	Average emissivity†	Total emissivity†
2000	2030	0.446	0.264
2200	2238	0.443	0.285
2400	2447	0.440	0.304
2600	2660	0.437	0.320
*2800	2874	0.434	0.334
3000	3092	0.432	0.347

Colour temperature. The temperature of a black body having the same colour as the tungsten source.

Average emissivity. The ratio of tungsten brightness to that of a black body at the same temperature.

Total emissivity. The ratio of the total energy radiated compared with that of a black-body at the same temperature.

* 100 watt gas-filled lamp.

LIGHT

PHOTOMETRIC UNITS

The candela (cd), the unit of luminous intensity, has been adopted as a basic unit of SI (for definition see p. 12).

The lumen (lm) is a unit of flux of light energy. It is the flux, per unit solid angle, from a source of unit intensity.

The lux (lx) is a unit of illumination, and is the illumination produced by a flux of 1 lumen on a surface of area 1m^2.

The nit, the unit of luminance, is defined as the luminous intensity of a surface per unit projected area, which is emitting 1 lumen per square metre per steradian, i.e. 1cd m^{-2}. The CGS unit (i.e. an emission of 1cd cm^{-2}) is the stilb (sb).

In the case of a uniformly diffusing surface (one which appears equally bright from whatever direction it is viewed) an alternative unit of luminance may be defined as the luminance of a surface of which unit area emits 1 lumen. In the metric system this is the Apostilb (1 lumen per m^2) and in the British system the foot-lambert (1 lumen per ft^2). One stilb equals 1000π apostilb.

Relative sensitivity of the eye to light of equal intensity

Wavelength (nm)	400	450	500	550	555	600	650	700	750
Sensitivity	0.04	3.8	32.3	99.5	100	63.1	10.7	0.41	0.01

The power equivalent of 1 lumen at maximum sensitivity (555 nm) = 1.47×10^{-3} W.

Light flux equivalent of 1 watt at maximum sensitivity = 680 lm.

EMISSION SPECTRA

The primary wavelength standard is that of radiation, in vacuum, corresponding to the transition between the levels $2p_{10}$ and $5d_5$ of the krypton-86 atom. This is 605.780 21 nm, (605.612 47 nm in dry air at 15°C, 101 325 N m⁻²).

A number of secondary standards are available, of slightly less precision, among them being the former primary standard cadmium-114 at 644.0248 nm (643.8469 nm in air). Others include mercury-198 at 546.2270 nm and 435.9562 nm (546.0752 and 435.8336 nm in air).

The principal lines of a few common substances are given below, all wavelengths in dry air at 15°C, 101 325 N m⁻² pressure. In general spectra are too complicated for abbreviated tables to be of use and reference should be made to standard treatises (see Preface). Wavelengths in the optical region have customarily been measured in ångströms (1 Å = 10^{-10} m = 10^{-1} nm), which are retained by SI for a limited time.

For the dispersion of dry air, 15°C, standard pressure, see page 69.

Flame Spectra. Principal lines (nm)

Sodium chloride.
D_1 589.5932
D_2 588.9965

Potassium chloride.
769.901
766.494
404.722
404.416

Strontium chloride.
Bands in red 675—623
460.7342

Lithium chloride.
670.7843

Fine structure of the mercury green line.

The mercury green line λ546.075 has components at

$$+215, +129, +84, 0, -70, -99, -235, \times 10^{-4} \text{ nm}.$$

with respect to the mean wave-length.

Discharge Tubes

Wavelengths in nm, dry air, 15°C, standard pressure.

Helium.	Hydrogen.	Argon.
706.520	Many-lined spectrum	706.7218
667.8149	and	6 96.5431
587.563	Hα **656.2784**	675.2831
501.5678	Hβ **486.1327**	603.2127
492.1926	Hγ 434.0466	565.0708
471.3147	Hδ 410.1736	549.5875
447.1480		**470.2317**
438.7930	**Neon.**	**462.8445**
414.3759		**459.6096**
412.0820	724.5166 540.0562	452.2325
402.6192	717.3938 535.802	451.0733
388.8645	650.6528 534.1096	**434.5168**
	640.2246 533.0779	433.535
Mercury.	638.2991 478.893	**433.3561**
1013.977	626.6495 471.534	**430.0100**
690.752	614.3062 470.886	**427.2169**
623.435	596.544 470.439	**426.6286**
579.065	**588.1895** 457.586	**425.9362**
576.960	**585.2488** 453.776	425.1184
546.073		**420.0678**
496.032	**He—Ne laser.**	**419.8316**
491.600	632.8164	**419.1027**
435.834		419.0714
434.750	**Cadmium.**	418.1884
433.923	**643.846 96**	416.4180
407.782	**508.582 30**	**415.8591**
404.656	**479.991 39**	**404.4419**
	467.815 04	**394.8980**
	466.235 25	

The more prominent lines are in bold type.

The following spectra are associated with leaks, impurities, etc., commonly met with in discharge tubes:

Carbon dioxide, monoxide. Numerous bands, shaded to violet. Principal heads: 662.0, 607.9, 561.6, 519.8, 483.5, 451.1.

Hydrocarbon vapours. Swan bands (blue cone of coal-gas flame). Multiple headed bands beginning 563.6, 516.5, 473.7, 438.0.

Hydrogen. Easily recognised by Hα and Hβ (see above).

Mercury. From diffusion pump or contaminated Al electrodes. Most persistent line, 546.1.

Nitrogen. Spark. Doublets at

> 594.2, 593.2, 568.0, 566.7, 500.5, 500.0.

> Tube. Bands at

> 678.7, 670.3, 662.2, 654.3, 646.7, 639.3, 632.1, 625.1, 606.8, 601.2, 595.7, 590.4, 585.3, 580.3, 575.4.

Oxygen. Spark. Chief lines, 6158, 5330.

> Tube. Diffuse bands, 564—555, 530—520.

Series Data

Hydrogen

$$\sigma = R(1/n_1^2 - 1/n_2^2).$$

R = Rydberg constant = $10\,967\,758\,\text{m}^{-1}$.

R_∞ = $10\,973\,731\,\text{m}^{-1}$.

n_1	n_2	Series.
1	2, 3, 4, ...	Lyman.
2	3, 4, 5, ...	Balmer.
3	4, 5, 6, ...	Paschen.
4	5, 6, 7, ...	Brackett.
5	6, 7, 8, ...	Pfund.

Singly ionized helium

$$\sigma = 4 \times 10\,972\,227\ (1/n_1^2 - 1/n_2^2)\,\text{m}^{-1}.$$

n_1	n_2	Series.
2	3, 4, 5, ...	Lyman.
3	4, 5, 6, ...	'4686'.
4	5, 6, 7, ...	Pickering.

In the past, spectroscopists have been accustomed to use wave-number (σ) (i.e. in the CGS system the number of waves in 1 cm in vacuo, in SI the number of waves in 1 m in vacuo) as a measure of frequency. For such expressions in CGS (cm^{-1}) units true frequency is then $\sigma\,(2.997\,925 \times 10^{10})\,\text{Hz}$, and σ is $10^8/\lambda$, with λ in Å. When σ is expressed in m^{-1} units, true frequency is $\sigma\,(2.997\,925 \times 10^8)\,\text{Hz}$ and σ is $10^9/\lambda$ if λ is in nm.

EMISSION SPECTRA

*Transverse **Zeeman** effect*

The wave-number changes for a line showing the normal effect are:

$$0, \pm 46.6858 \text{ m}^{-1} \text{ per weber per metre}^2.$$

Line (nm)	Zeeman type	Wave-length change (nm) per Wb m^{-2}
He 667.8	$\pm$ (0) 1:1	$\pm$ (0), 0.0209
Na 589.6	$\pm$ (2) 4:3	$\pm$ (0.0109), 0.0217
Na 589.0	$\pm$ (1) 3, 5:3	$\pm$ (0.0054), 0.0162, 0.0271
Ne 588.1	$\pm$ (0) (1) 8, 9, 10:6	$\pm$ (0) (0.0027), 0.0215, 0.0242, 0.0270
Ne 585.2	$\pm$ (0) 31:30	$\pm$ (0), 0.0165
Hg 579.0	$\pm$ (0) 1:1	$\pm$ (0), 0.0157
Cd 468.0	$\pm$ (0) 4:2	$\pm$ (0), 0.0205

The values in parenthesis denote components which are plane polarized with electric vector parallel to the direction of the applied magnetic field, the remainder being polarized perpendicularly to the field.

THE ELECTROMAGNETIC SPECTRUM

The limits of the sections are quite arbitrary, but in accordance with common usage. Wave-lengths are in metres.

Region.	Wave-length range.
Wireless telegraphy	
Long wave	10^3 to 10^4
Medium wave	10^2 to 10^3
Short wave	10 to 10^2
Residual waves } Far infrared	10^{-5} to 10^{-4}
Near infrared	7.5×10^{-7} to 10^{-5}
Visual	
Red	6.5×10^{-7} to 7.5×10^{-7}
Orange	5.9×10^{-7} to 6.5×10^{-7}
Yellow	5.3×10^{-7} to 5.9×10^{-7}
Green	4.9×10^{-7} to 5.3×10^{-7}
Visual (*cont.*)	
Blue	4.2×10^{-7} to 4.9×10^{-7}
Violet	4.0×10^{-7} to 4.2×10^{-7}
Ultra-violet	1.8×10^{-7} to 4.0×10^{-7}
Schumann } Lyman	10^{-8} to 1.8×10^{-7}
X-rays	
Soft	10^{-10} to 2.0×10^{-9}
Hard	10^{-11} to 10^{-10}
Gamma rays	5×10^{-12} to 5×10^{-11}
Cosmic ray components	10^{-14}

Isotropic solids

The index of refraction (n) is given for room temperature for the sodium D lines λ 589.3.

Amber	1.546	Potassium bromide	1.5600
Canada balsam	1.530	Quartz fused (18°C)	1.45843
Cellulose acetate	1.49–1.50	Sodium chloride	1.5443
Diamond	2.4173	Strontium titanate	2.407
Fluorite CaF_2	1.434	Sylvine KCl	1.4904
Gelatine	1.530	Zinc sulphide	2.370

Some representative glasses

	H_α 656.3 n_C	Na 589.3 n_D	H_β 486.1 n_F	H_γ 434.0 n_C'	$\dfrac{n_D - 1}{n_F - n_C}$
Borosilicate crown	1.5136	1.5160	1.5217	1.5262	63.7
Dense barium crown	1.5852	1.5881	1.5949	1.6003	60.6
Light flint	1.5746	1.5787	1.5888	1.5973	40.8
Extra dense flint	1.6415	1.6469	1.6607	1.6724	33.7

Liquids λ 589.3 Na; temperature 20°C.

		For 1°C rise subtract			For 1°C rise subtract
Acetone	1.3585	0.00053	Ethanol	1.3618	0.00040
Aniline	1.5863	0.00054	Ether	1.3525	0.00059
Benzene	1.5014	0.00064	Glycerol	1.4730	–
Carbon disulphide	1.6279	0.00079	Methanol	1.3292	0.00036
Carbon tetrachloride	1.4607	0.00050	Monobrom-naphth	1.6588	–
Cinnam-aldehyde	1.6195	–	Oil Cedarwood	1.516	–
Chloroform	1.4453	0.00055	Olive	1.47	–
			Paraffin	1.43	–
			Turpentine	1.48	–

Water

	10°C	15°C	20°C	25°C	30°C
656.3 H_α	1.3318	.3315	.33115	.3307	.3302
589.3 Na	.33374	.33341	.33299	.33252	.33192
486.1 H_β	.3378	.3375	.33714	.3366	.3360
434.0 H_γ	.3411	.3408	.34055	.3398	.3392

Solutions

Aqueous solutions; λ589.3 Na; temperature 18°C.

Refractive index of solution

$$= 1.3332\,(Ac - Bc^2).$$

c = g of salt per 100 cm³ solution.

Salt	c	A	B
Lead nitrate	0–30	0.00116	0.000 001 1
Potassium nitrate	0–20	0.00096	0.000 004 9
Potassium chloride	0–20	0.00138	0.000 007 2
Potassium tartrate	0–30	0.00136	0.000 005 1
Sodium chloride	0–20	0.00176	0.000 008 8

Gases For λ589.3 Na, 0°C and standard pressure.

Acetylene	1.000 606	Hydrogen chloride	1.000 447
Air (no CO_2)	1.000 292	Hydrogen sulphide	1.000 637
Ammonia	1.000 376	Methane	1.000 444
Argon	1.000 282	Neon	1.000 067
Carbon dioxide	1.000 450	Nitric oxide	1.000 297
Carbon monoxide	1.000 335	Nitrogen	1.000 298
Chlorine	1.000 781	Nitrous oxide	1.000 512
Helium	1.000 035	Oxygen	1.000 271
Hydrogen	1.000 139	Sulphur dioxide	1.000 686

Refractive index of air, 15°C, standard pressure.

$(n - 1) \times 10^8 = 27\,259.9 + 153.58\sigma^2 + 1.318\sigma^4 + 0.0155\sigma^6$

where $\sigma = 10^{-6}$ (wave number m^{-1}).

UNIAXIAL CRYSTALS. FOR λ589.3 Na

	Ordinary ray	Extraordinary ray
Beryl (Emerald)	1.581	1.575
Calcite	1.6591	1.4868
Corundum (Sapphire, Ruby)	1.7686	1.7604
Ice	1.3091	1.3104
Magnesium fluoride	1.378	1.390
Tourmaline	1.669	1.638

	656.3 H$_\alpha$	589.3 Na	486.1 H$_\beta$	434.0 H$_\gamma$
Quartz				
Ordinary ray	1.541 90	1.544 25	1.549 68	1.553 96
Extraordinary ray	1.550 93	1.553 36	1.558 98	1.563 41

BIAXIAL CRYSTALS
The three principal indices for λ589.3 Na

Gypsum	1.5205	1.5226	1.5296
Mica	1.561	1.590	1.594
Aragonite	1.530	1.681	1.686
Barite	1.637	1.638	1.649
Potassium nitrate	1.335	1.506	1.507
Cane sugar	1.537	1.565	1.571
Cryolite	—	1.364	—

ROTATORY POWERS

p = no. of grams of substance per 100 g solution.

Substance	Solvent	Specific rotation $[\alpha]_D^{20}$ Degrees	p
Sucrose (cane sugar)	Water	$66.412 + 0.0127p$ $- 0.000\,377p^2$	0–50
Dextrose (glucose)	Water	$52.50\ + 0.0188p$ $+ 0.000\,517p^2$	0–35
Invert sugar	Water	$-19.447 - 0.0607p$ $+ 0.000\,221p^2$	10–65
Tartaric acid	Water	$14.83\ \ - 0.146p$	0–50
Turpentine	Alcohol	$-37.0 - 0.004\,82p$ $- 0.000\,13p^2$	0–90
Camphor	Alcohol	$40.9 + 0.135p$ $54.4 - 0.135p$	10–50 50–90

Rotation for quartz Degrees per mm; temperature 20°C

	656.3 H$_\alpha$	589.3 Na	486.1 H$_\beta$	434.0 H$_\gamma$
Rotation	17.320	21.724	32.761	41.924

(For 589.3 Na add 0.003 per 1°C rise.)

REFLECTING POWER OF METALS

For thick layers; percentage reflected at normal incidence.

	Wave-length nm					
	800	700	600	500	400	300
Silver	97	96	94	91	87	*10
Gold	95	92	84	47	28	32
Nickel	70	69	65	61	53	44
Steel	58	58	55	54	50	37
Platinum	70	69	64	58	48	40
Copper	89	83	72	44	31	25
Speculum	69	67	64	63	55	41

*Silver has a very low reflecting power in the region 316.

Reflecting power, for normal incidence, of transparent substances in air (single reflection).

Refractive index										
1.5	1.6	1.7	1.8	1.9	2.0	2.1	2.2	2.3	2.4	2.5
$R\%$ 4.0	5.3	6.7	8.2	9.6	11.0	12.6	14.0	15.5	17.0	18.4

MAGNETISM AND ELECTRICITY

ELECTROMOTIVE FORCES

Primary Cells

Weston standard cell. e.m.f. at 20°C = 1.01861V; over the range 0° to 40°C, e.m.f. at t°C

$$= 1.01861 - 37 \times 10^{-6} (t - 20) - 11 \times 10^{-7} (t - 20)^2.$$

Clark standard cell. e.m.f. at 15°C = 1.4332V; over the range 0° to 30°C, e.m.f. at t°C

$$= 1.4332 - 119 \times 10^{-5} (t - 15) - 0.7 \times 10^{-5} (t - 15)^2.$$

The maximum current which may be taken from standard cells without perceptible polarization is of the order 5×10^{-5} A.

Common Primary Cells

Cell	Approx. open circuit e.m.f.	Elements
Lechlanche	1.5 volts	(Amalgamated zinc) (Saturated salammoniac soln.) (Manganese dioxide and carbon) (Carbon).
Daniell	1.1 volts	(Amalgamated zinc) (60 cm³ sulphuric acid per litre water) (Saturated copper sulphate soln.) (Copper).
Bichromate	2.0 volts	(Amalgamated zinc) (90 g $K_2Cr_2O_7$, 100 cm³ H_2SO_4, 900 cm³ H_2O) (Carbon).
Bunsen	1.9 volts	(Amalgamated zinc) (80 cm³ H_2SO_4 per litre water) (Conc. HNO_3) (Carbon)

MAGNETIC PROPERTIES OF FERROMAGNETIC SUBSTANCES

The table gives the magnetic flux density (B) Wb m^{-2}.

Permeability (μ) = B/H.

Substance	H (Am^{-1})						
	8	40	80	8×10^2	4×10^3	8×10^3	8×10^4
Cast iron (ann.)	—	—	0.14	0.52	0.9	1.0	1.6
Mild Steel (ann.)	—	0.7	1.0	1.6	1.7	1.8	2.2
Swedish iron (ann.)	—	0.1	0.6	1.5	1.7	1.8	2.2
Cobalt steel	—	—	0.96	0.94	0.87	0.73	0
Transformer iron	—	0.16	0.52	1.4	1.6	1.7	—
Permalloy	0.6	0.9	1.0	1.1	1.1	1.1	—
Nickel	—	—	—	0.43	0.55	0.60	—
Cobalt	—	—	—	—	0.60	1.0	-

Substance	Coercive force A m^{-1}	H A m^{-1}	Initial relative perm.	Maximum relative perm.	Steinmetz coeff.
Cast iron (ann.)	370	400	50	600	3.25
Mild steel (ann.)	32	40	350	14 000	0.750
Swedish iron (ann.)	64	80	—	6 400	0.500
Cobalt steel	16 000	—	—	—	—
Transformer iron	—	88	500	5 500	0.250
Permalloy	6.4	8	9000	85 000	—
Nickel	640	—	—	300	6.25
Cobalt	960	—	—	170	3.00

Hysteresis loss (J kg^{-1}) = (Steinmetz coeff.)(max. flux density)$^{1.6}$

MAGNETIC SUSCEPTIBILITY (χ_m)

The susceptibility of some common substances; the values are per kg at a temperature of 18°C. In every case the unit is 10^{-8}.

Solids

Aluminium	+ 0.81	Sulphur	− 0.62
Bismuth	− 1.70	Tin	+ 0.031
Cadmium	− 0.23	Tungsten	+ 0.35
Copper	− 0.107	Zinc	− 0.197
Diamond	− 0.62	Ebonite	+ 0.75
Gold	− 0.19	Glass (soda)	abt. − 1.25
Graphite	− 4.4	Quartz	− 0.58
Lead	− 0.15	Shellac	− 0.38
Platinum	+ 1.22	NaCl	− 0.63
Sodium	+ 0.75	KNO$_3$	− 0.41

(for semiconductors, see table p. 29)

MAGNETIC SUSCEPTIBILITY

Liquids

Water 20°C	− 0.905	Ethanol 18°C	− 1.00
Mercury 18°C	− 0.23	Methanol 18°C	− 0.93
Oxygen − 196°C	+ 330	Glycol 18°C	− 0.68
Carbon		Benzene 18°C	− 0.89
tetrachloride 18°C	− 0.54		

Gases 1 atm pressure, 20°C

Air	+ 30.4	Methane	− 3.1
Argon	− 0.59	Neon	− 0.41
Carbon dioxide	− 0.56	Nitric oxide	+ 60.2
Helium	− 0.59	Nitrogen	− 0.54
Hydrogen	− 2.48	Oxygen	+ 133.6

CHANGE OF RESISTANCE OF BISMUTH IN A TRANSVERSE MAGNETIC FIELD

Temperature 18°C

Field (10^2 A m^{-1})	0	2	4	6	8	10	13	16	19	22	25
Relative resistance	1.00	1.05	1.17	1.29	1.43	1.57	1.80	2.03	2.26	2.49	2.72

MAGNETO-OPTICAL ROTATION (FARADAY EFFECT)

At 20°C for 589.3 Na. Rotation per m path length per A m^{-1}.

Substance	Rotation (minutes)
Water	0.016 44
Alcohol	
ethyl	0.013 97
methyl	0.011 8
Benzene	0.037 3
Carbon disulphide	0.053 11
Quartz (perp. to axis)	0.020 91
Glass (extra dense flint)	0.111

In every case the rotation is clockwise for light travelling in the same direction as the lines of force.

APPROXIMATE VALUES
OF THE MAGNETIC ELEMENTS (1971)

Latitude	Longitude					
	8°W	6°W	4°W	2°W	0°	2°E
58°N	13.1 (71.1)	13.2 (70.7)	13.2 (70.5)	13.3 (70.4)	13.3 (70.3)	13.4 (70.2)
56°N	13.7 (69.6)	13.8 (69.4)	13.8 (69.3)	13.9 (69.1)	14.0 (69.0)	14.0 (68.9)
54°N	14.4 (68.3)	14.4 (68.2)	14.4 (68.0)	14.5 (67.9)	14.6 (67.7)	14.7 (67.5)
52°N	15.0 (67.1)	15.1 (66.9)	15.1 (66.8)	15.2 (66.6)	15.2 (66.4)	15.3 (66.2)
50°N	15.6 (65.9)	15.7 (65.7)	15.7 (65.5)	15.8 (65.3)	15.9 (65.1)	16.0 (64.9)

The upper value gives the horizontal force (A m^{-1}).
The lower bracketed value gives the dip (degrees).

Declination at Greenwich   6° 19′ W.

	Latitude	Longitude
North magnetic pole	76°N	100°W
South magnetic pole	66°S	140°E

RESISTIVITY (ρ) OF SOME COMMON SUBSTANCES

Temperature 20°C; temperature coefficient at 20°C. The values are in $10^{-8}\ \Omega$ m.

Metals

	$\rho(10^{-8}\Omega$ m)		α
Aluminium	2.67		0.004 5
Antimony	44		0.005 1
Bismuth	117		0.004 6
Brass	abt. 8	abt.	0.001 5
Bronze (phosphor)	abt. 8	abt.	0.003 5
Cobalt	6.4		0.006 6
Constantan	49		0.000 01
Copper	1.72		0.004 2

75

RESISTIVITY

Metals (cont.)

	$\rho(10^{-8}\Omega\,m)$	α
Be-copper	6.8–7.4	0.0013
German silver	abt. 28	abt. 0.0003
Gold	2.20	0.0040
Iridium	5.2	0.0045
Iron	10.3	0.0065
Lead	20.63	0.0042
Magnesium	4.24	0.0043
Manganin	abt. 44	abt. 0.00001
Mercury	95.93	0.0089
Molybdenum	5.7	0.0044
Nichrome	abt. 100	abt. 0.0004
Nickel	6.94	0.0070
Palladium	10.7	0.0039
Platinoid	abt. 38	abt. 0.00028
Platinum	10.5	0.0039
Silver	1.63	0.0038
Steels		
Hardened	abt. 45	abt. 0.0015
Invar	78	0.002
Mild	abt. 15	abt. 0.0033
Piano wire	abt. 12	abt. 0.0032
Tantalum	13.4	0.0033
Tin	11.4	0.0044
Tungsten	5.5	0.0050
Zinc	5.92	0.0038

(for semiconductors, see table on p. 29)

Insulators

Temperature approximately 18°C

	$\Omega\,m$		$\Omega\,m$
Amber	5×10^{14}	Quartz	
Bakelite (pure, no filler)	2×10^{14}	(parallel to axis)	1×10^{12}
Celluloid	2×10^{8}	(perp. to axis)	3×10^{14}
Ebonite	2×10^{13}	Fuzed quartz	5×10^{16}
Fibre	2×10^{8}	(600°C)	6×10^{5}
Plate glass	2×10^{11}	Sealing wax	1×10^{13}
Mica (clear)	5×10^{14}	Shellac	1×10^{14}
Paraffin wax	3×10^{16}	Slate	1×10^{6}
Unglazed porcelain	3×10^{12}	Sulphur	1×10^{15}

The resistance of most insulators decreases rapidly with temperature, and at 30°C may be only half that given above.

(for plastics, see table on pp. 38, 39)

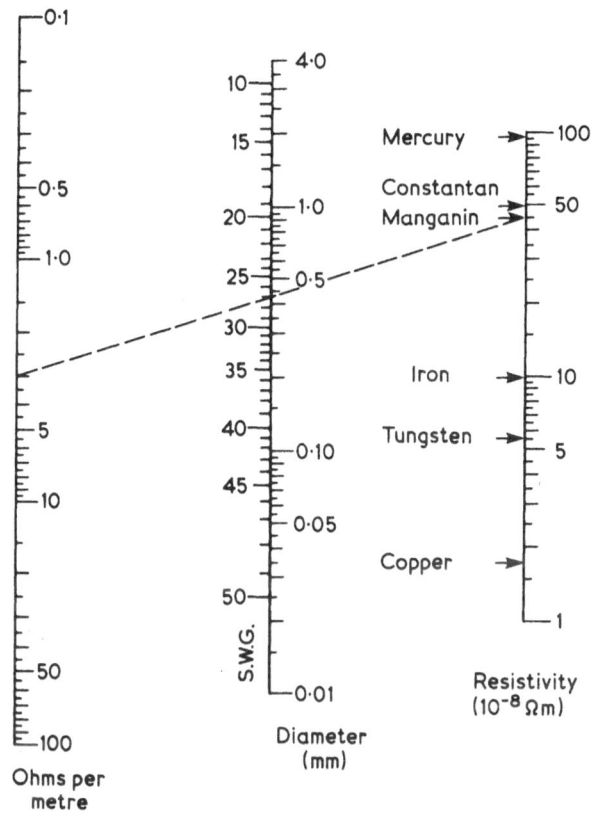

Nomogram 6. Rapid calculation of wire resistances.
A straight edge placed across the diagram passes through three inter-related quantities. Thus the resistance of 27-gauge Manganin is 3 ohms per metre.

THERMO-ELECTRIC E.M.F. OF SOME COMMON METALS AND ALLOYS, AGAINST LEAD (Pb)

The values are approximate, since purity and structure are important factors. Cold junction temperature, 0°C

$$\text{E.M.F. (microvolts)} = At + Bt^2.$$

A positive e.m.f. indicates that at the hot junction the current is *out of* the lead (Pb). For a junction composed of any two metals take the difference of their e.m.f.'s (note signs).

Metal	A	B	Temp. range, °C
Aluminium	−0.50	0.00086	0 to 200
Bismuth	−74	−	0 to 100
Brass	0.14	0.0026	0 to 100
Constantan	−38.1	−0.045	0 to 400
Copper	2.76	0.006	0 to 100
German silver	−10.9	−0.016	−200 to 100
Gold	2.9	0.0046	0 to 200
Iridium	2.4	−0.0014	−80 to 100
Iron	16.7	−0.015	−230 to 100
Mercury	−8.8	−0.016	0 to 200
Nichrome	25	−	0 to 420
Nickel	−16.3	−0.027	−200 to 100
Platinum	−3.0	−0.016	−200 to 300
Silver	3.0	0.0034	−200 to 100
Tin	−0.17	0.0009	0 to 200
Tungsten	1.59	0.017	0 to 100
Zinc	3.05	−0.02	0 to 100

STANDARD WIRE GAUGE

The values given are diameters in mm

		10	20	30	40	50
0	8.23	3.25	0.914	0.315	0.122	0.025
1	7.62	2.95	.813	.295	.112	
2	7.01	2.64	.711	.274	.102	
3	6.40	2.34	.610	.254	.091	
4	5.89	2.03	.559	.234	.081	
5	5.39	1.83	.508	.213	.071	
6	4.88	1.63	.457	.193	.061	
7	4.47	1.42	.417	.173	.051	
8	4.06	1.22	.376	.152	.041	
9	3.66	1.02	.345	.132	.031	

SPARKING DISTANCES, SPHERICAL ELECTRODES

Diameter of electrodes, 2.0 cm

Gap mm	Volts	Gap mm	Volts	Gap mm	Volts
2	8.5×10^3	7	23.7×10^3	20	52.1×10^3
3	11.8	8	26.5	25	58.5
4	14.9	9	29.0	30	64.2
5	17.9	10	32.0	35	68.6
6	20.9	15	42.4	40	71.9

ELECTROCHEMICAL EQUIVALENTS ($mg\,C^{-1}$)

Silver	1.117 98	Hydrogen	0.010 45
Copper	0.329 27	Oxygen	0.082 91
Lead	1.073 67	Mercury	2.078 94

Electrochemical equivalent per unit equivalent weight
$$= 0.010\,364\,12\,mg\,C^{-1}$$
(thus ECE mercury = 0.010 364 12 (at wt Hg)/Valency = 2.078 94.
Faraday constant per kg equivalent $9.648\,670 \times 10^7\,C$.

RESISTOR/CAPACITOR COLOUR CODE

Resistor Band No	Resistor Function	Black	Brown	Red	Orange	Yellow	Green	Blue	Violet	Grey	White	Silver	Gold	Capacitor Band No	Capacitor Function
		−90	−90	−30	−80	−150	−220	−330	−470	−750	+30	−	+100	1	Temp. Coeff. 10^{-6} °C^{-1}
1	First Digit	0	1	2	3	4	5	6	7	8	9	−	−	2	First Digit
2	Second Digit	0	1	2	3	4	5	6	7	8	9	−	−	3	Second Digit
3	Multiplier	1	10	10^2	10^3	10^4	10^5	10^6	10^7	10^8	−	10^{-2}	10^{-1}		
		1	10	10^2	10^3	10^4	−	−	−	10^{-2}	10^{-1}	10^{-2}	10^{-1}	4	Multiplier
4	Tolerance %	±20	±1	±2	±3	−	±5	−	−	−	−	±10	±5		
		±20	±1	±2	±3	−	±5	−	−	−	±10	±10	−	5	Tolerance % (above 10 pF)
		±2	±0.1	−	−	−	±0.5	−	−	±0.25	±1.0	−	−	5	Tolerance pF (10 pF and below)

RESISTOR/CAPACITOR COLOUR CODE

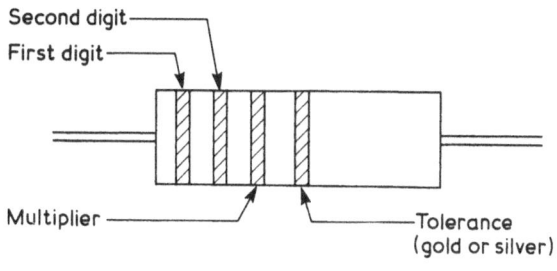

RESISTOR

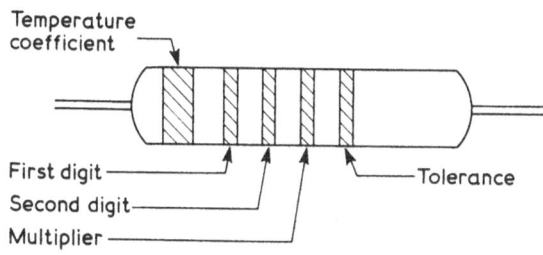

CAPACITOR

The resistor code is fairly general, the capacitor code is subject to variation.

CONDUCTIVITY (x), AND EQUIVALENT CONDUCTIVITY OF AQUEOUS SOLUTIONS

Conductivity values of solutions, $\Omega^{-1}\,m^{-1}$. Conductivity of good distilled water, 10^{-4}.

Standard calibration solutions

Solute	5°C	10°C	15°C	20°C	25°C
NaCl*	15.55	17.79	20.14	22.60	25.13
KCl**	7.414	8.319	9.252	10.207	11.180
1/10 KCl	0.822	0.933	1.048	1.167	1.288
1/100 KCl	0.090	0.102	0.115	0.128	0.141

* Saturated at all temperatures.

** Normal, $74.59\,kg\,m^{-3}$ at 18°C

CONDUCTIVITY OF AQUEOUS SOLUTIONS

Equivalent Conductivities, 18°C (Ω^{-1})

kg equivalent per m^3	KCl	KNO$_3$	NaCl	AgNO$_3$	1/2CuSO$_4$	HCl	1/2H$_2$SO$_4$	KOH
0	130.1	126.5	109.0	115.8	115.0	378	383	238
0.0001	129.1	125.5	108.1	115.0	109.9	377	380	237
0.001	127.3	123.6	106.5	113.1	98.5	376	361	234
0.01	122.4	118.2	102.0	107.8	71.7	369	308	228
0.1	112.0	104.8	92.0	94.3	43.8	351	225	213

Stronger solutions, 18°C

				Conductivity (Ω^{-1})				
p	KCl	c	KNO$_3$	c	NaCl	c	AgNO$_3$	c
5	0.069	2	0.046	1.6	0.067	2.2	0.026	2.2
10	0.136	1.9	0.094	–	0.121	2.1	0.048	2.2
15	0.202	1.8	0.125	–	0.164	2.1	0.068	2.2
20	0.268	1.7	0.144	–	0.196	2.2	0.087	2.1

p	CuSO$_4$	c	HCl	c	H$_2$SO$_4$	c	KOH	c
5	0.019	2.2	0.395	1.58	0.209	1.21	0.172	1.9
10	0.032	2.2	0.630	1.56	0.392	1.28	0.315	1.9
15	0.042	2.3	0.745	1.55	0.543	1.36	0.425	1.9
20	–	–	0.762	1.54	0.653	1.45	0.499	2.0

p = % by weight of anhydrous substance.

c = % correction added per 1°C rise of temperature.

DIELECTRIC CONSTANT (ε_r, $\varepsilon/\varepsilon_0$)
AND DIELECTRIC STRENGTH

Solids

	ε_r	S		ε_r	S
Amber	2.8		Porcelain	5.6	
Chlornapth	2.8		Quartz		
Ebonite	2.8	30—110	(perp. to		
Glass			axis)	4.5	—
Flint	7—10	30—150	(par. to axis)	4.6	—
Crown, plate	5—7	30—150	fused	3.7	
Mica	6	80—200/.05	Rubber	2.2	16—40
Paper	2	3.6/.2	Shellac	3.1	—
(paraffined)		40—60	Sulphur	4—4.2	—
Paraffin wax	2.2	15—50	Wood	2—8	0.4—0.6/25

ε_r, dielectric constant; S, dielectric strength in kilovolt per
  mm/thickness of sample in mm.

Liquids Temperature, 18°C

	ε_r		ε_r
Ethanol	26	Oil	
Methanol	32	olive	3.1
Aniline	7.3	paraffin	4.7
Benzene	2.3	transformer	2.2
Carbon disulphide	2.6	Turpentine	2.3
Carbon tetrachloride	2.2	Vaseline	1.9
		Water	81

Gases Temperature, 0°C Pressure, 760 mmHg

	ε_r	F		ε_r	F
Air	1.000 59	1.0	Hydrogen	1.000 27	0.65
Ammonia	1.007 20	1.0	Methane	1.000 94	—
Carbon monoxide	1.000 65	—	Nitrogen	1.000 60	1.15
Carbon dioxide	1.000 99	0.95	Oxygen	1.000 53	0.85
Helium	1.000 07		Sulphur dioxide	1.009 50	0.30

ε , dielectric constant; F, Rupture voltage/air rupture voltage.

Accelerating field			Shortest e.m. wavelength generated	de Broglie wavelength	Range in Al
kV	Vel./c	m/m_0	nm	pm	mg cm^{-2}
10	0.195	1.02	12.4	12.26	0.16
20	0.272	1.04	6.2	8.67	0.68
40	0.374	1.08	3.1	6.13	2.60
60	0.447	1.12	2.07	5.01	5.5
80	0.504	1.16	1.55	4.34	9.2
100	0.549	1.20	1.24	3.88	13.5
500	0.863	1.98	0.248	1.73	162
1000	0.941	2.96	0.124	1.23	412
1500	0.967	3.94	0.083	1.00	680
2000	0.979	4.91	0.062	0.87	950

Compton wavelength of electron 2.4263 pm

c = velocity of light

m_0 = rest mass

X-RAY DATA

Effective grating space of calcite 303.534 pm (303.575 true)
Grating space of rocksalt 281.969 pm

K-series   The wavelengths are given in pm units (10^{-2} Å)

Element No.	Line			
	α_2	α_1	β_1	β_2
13 Aluminium	834.152	833.907	798.1	
24 Chromium	229.356	228.967	208.468	207.09
27 Cobalt	179.282	178.891	162.072	160.898
29 Copper	154.4361	154.0516	139.217	138.104
26 Iron	193.9942	193.5998	175.6572	174.433
82 Lead	17.0287	16.5366	14.5964	14.191 14.212
25 Manganese	210.576	210.177	191.007	189.711
42 Molybdenum	71.3551	70.9267	63.2259	62.0956
28 Nickel	166.178	165.786	150.009	148.863
78 Platinum	19.0374	18.5506	16.3666	15.9187 15.9376
47 Silver	56.3780	55.9368	49.702	48.702
74 Tungsten	21.3820	20.8998	18.4340	17.9401 17.9576
30 Zinc	143.895	143.508	129.517	128.367

L-series

Line	Substance. Molybdenum	Platinum	Tungsten
1	615.06	149.94	167.84
α_2	541.412	132.423	148.739
α_1	540.630	131.30	147.635
η	584.72	124.28	142.10
β_4 }	504.86	114.217	130.143
β_6 }		114.332	128.96
β_1	517.684	111.985	128.177
β_3	501.32	110.389	126.248
β_2	492.31	110.197	124.455
γ_1	472.57	95.793	109.853
γ_2 }	437.98	93.422	106.804
γ_3 }		92.787	106.202

ATOMIC AND MOLECULAR CONSTANTS

Gas	Mean free path s.t.p. 10^{-8} m	Root mean square vel. s.t.p. 10^2m s^{-1}	No. impacts per second s.t.p. 10^9s^{-1}	Effective diameter (visc.) 10^{-10}m	Nuclear distance (spect.) 10^{-10} m
Argon	6.4	4.1	6.0	3.5	–
Helium	17.7	13.1	6.8	2.1	–
Hydrogen	11.2	18.4	15.1	2.6	0.75
Neon	12.6	5.8	4.2	2.5	–
Nitrogen	6.0	4.9	7.6	3.6	1.09
Oxygen	6.4	4.6	6.6	3.5	1.20

The mean free path for a given gas is inversely proportional to the density.

RADIOACTIVITY CONSTANTS OF NATURALLY OCCURRING RADIOELEMENTS

Substance	Species	Half-life T	Energies (MeV)		
			α	β	γ*
Carbon	14 C	5568 y	–	0.155	–
Potassium	40 K	12.6×10^8 y	–	1.33	1.46
Rubidium	87 Rb	6.3×10^{10} y	–	0.27	–
Samarium	147 Sm	1.3×10^{11} y	2.1	–	–
Lutetium	176 Lu	7.5×10^{10} y	–	0.40	0.09; 0.18; 0.27
Thorium	232 Th	1.39×10^{10} y	4.0	–	–
Mesothorium I	228 Ra	6.7 y	–	0.053	–
Mesothorium II	228 Ac	6.13 h	–	(a)	(a)
Radiothorium	228 Th	1.90 y	5.42; 5.34	–	0.084
Thorium X	224 Ra	3.64 d	5.68; 5.45; 5.19	–	–
Th.eman.	220 Em	54.5 s	6.28	–	–
Thorium A	216 Po	0.158 s	6.74	–	–
Thorium B	212 Pb	10.6 h	–	0.36; 0.58	0.24; 0.30
Thorium C	212 Bi	60.5 min	6.09; 6.05	2.25	0.04
Thorium C'	212 Po	3.0×10^{-7} s	8.78	–	–
Thorium C''	208 Tl	3.1 min	–	1.79	2.62
Lead	208 Pb	stable	–	–	–
Act.uran.	235 U	7.1×10^8 y	4.58; 4.47; 4.40	–	–
Uranium Y	231 Th	25 h	–	0.30; 0.22; 0.09	–
Protoactinium	231 Pa	3.43×10^4 y	5.0; 4.8	–	(a)
Radioactinium	227 Th	18.6 d	6.0; 5.7	–	0.05; 0.12; 0.28
Actinium X	223 Ra	11.7 d	5.7; 5.6; 5.5	–	0.18; 0.27; 0.34
Act.eman.	219 Em	3.92 s	6.8; 6.6; 6.4	–	0.12; 0.27; 0.59
Actinium A	215 Po	1.83×10^{-3} s	7.37	–	–
Actinium B	211 Pb	36.1 min	–	1.39; 0.50	(a)

Actinium C'	211 Bi	2.16 min	6.6; 6.3	—	0.35
Actinium C''	211 Po	0.52 s	7.4; 6.9; 6.6; 6.3	—	—
Actinium C''	207 Tl	4.8 min	—	1.45	0.87
Lead	207 Pb	stable	—	—	—
Uranium I	238 U	4.5×10^9 y	4.2	—	0.05
Uranium X$_1$	234 Th	24.1 d	—	0.20; 0.10	0.09; 0.18
Uranium X$_2$	234 Pa	70 s	—	2.32; 1.50	0.4; 0.8
Uranium II	234 U	2.5×10^5 y	4.76	—	0.05; 0.09; 0.12
Ionium	230 Th	8.1×10^4 y	4.68; 4.6	—	0.07; 0.14; 0.23
Radium	226 Ra	1620 y	4.78	—	0.187
Radium eman.	222 Em	3.825 d	5.49	—	—
Radium A	218 Po	3.05 min	6.00	—	—
Radium B	214 Pb	26.8 min	—	0.70	0.24; 0.30; 0.35
Radium C	214 Bi	19.7 min	5.51; 5.45	3.17; 1.65	(a)
Radium C'	214 Po	1.6×10^{-4} s	7.68; 9.07	—	—
Radium C''	210 Tl	1.32 min	—	1.9	—
Radium D	210 Pb	22 y	—	0.02; 0.06	0.047
Radium E	210 Bi	5.0 d	—	1.17	—
Radium F	210 Po	138.3 d	5.30	—	—
Lead	206 Pb	stable	—	—	—

* Gamma-ray emission is from daughter elements.

The value given for β-particles shows the upper energy limit for spectra.

(a) indicates a spectrum too complex for abbreviation.

$$\text{Transformation constant } (\lambda) = \frac{0.693}{T}. \qquad \text{Average life} = \frac{1}{\lambda}.$$

$$T = \log_e 2/\lambda$$
$$ = 0.693/\lambda.$$

MASS ABSORPTION COEFFICIENT FOR X- AND GAMMA-RAYS

Mass absorption coefficient = Absorption coefficient/density $= \kappa/\rho$.

If a beam of initial intensity I_0 traverses thickness x(m) of substance, then final intensity $I = I_0 e^{-kx}$.

The unit in all cases is $kg^{-1} m^2$.

Wave-length pm	E	Aluminium	Copper	Carbon	Iron	Lead	Silver	Air	Water
	MeV				Substance				
0.1	12.4	0.00226	0.00318	0.00175	0.0030	0.0054	0.00405	0.00178	0.0021
0.2	6.2	0.00266	0.00306	0.0024	0.0030	0.0044	0.00358	0.00250	0.0026
0.5	2.5	0.00403	0.00385	0.0039	0.0038	0.0044	0.00385	0.00396	0.0043
1.0	1.24	0.00548	0.0057	0.0059	0.0054	0.0058	0.00480	0.0058	0.0063
	keV								
5	248	0.0111	0.0149	0.0113	0.0141	0.0577	0.0294	0.0113	0.0128
10	124	0.0153	0.0331	0.0144	0.0274	0.331	0.113	0.0145	0.0160
15	83	0.0193	0.0723	0.0158	0.0566	0.233	0.271	0.0165	0.0181
20	62	0.0258	0.150	0.0175	0.110	0.483	0.548	0.0184	0.0202
40	31	0.100	0.984	0.0244	0.726	3.00	3.61	0.0323	0.0345
60	20.6	0.300	3.09	0.0425	2.29	8.84	1.67	0.0700	0.0700
80	15.5	0.690	7.23	0.0800	5.29	14.2	3.94	0.132	0.139
100	12.4	1.31	12.9	0.1436	10.0	7.5	7.30	0.254	0.250
120	9.7	2.21	22.4	0.227	16.5	12.4	11.70	0.460	0.416
140	8.9	3.50	3.84	0.350	25.4	18.5	17.5	0.903	0.688
160	7.8	5.16	5.55	0.503	36.1	25.8	23.9	—	0.994
180	6.9	7.26	7.69	0.711	5.95	35.3	31.2	—	1.45
200	6.2	10.0	10.30	0.983	7.82	46	40.0	—	2.10

SOUND

VELOCITY OF SOUND

All values in metres per second
Solids, (Rod-shaped) 20°C

Aluminium	5100	Silver	2600
Brass	3400	Zinc	3700
Copper	3600	Brick	3700
Iron	5000	Slate	4500
(steel)	5100	Glass	5000
Nickel	4900	Wood	3000–4000
Platinum	2700		

Liquids, 20°C

Methanol	1120
Water	1483
(0–100°C) at t°C	
$1403 + 4.2t - 0.028t^2$	
Sea water (0–25°C)	
$1449 + 3.6t$	

Gases. s.t.p.

Air (dry)	331.46
at t°C	$331.46 + 0.61t$
Carbon dioxide	258
Methane	430
Hydrogen	1286
Oxygen	315
Helium	972

THE MEASUREMENT OF NOISE

The logarithmic scale for the comparison of sound energies is graduated in units of 1 'bel', subdivided for convenience into 10 decibels. Thus for two sounds having energies E_1 and E_2, $10 \log_{10} E_1/E_2$ = relative intensity in decibels. An increase of one decibel represents, therefore, an increase of 26 per cent. This is roughly the smallest change which the average person can appreciate, so that the logarithmic scale, when based on the faintest sound which is perceptible to a normal ear, is very appropriate for the description of 'loudness' of noise. For example, the noise of a pneumatic road drill at 20 ft is 90 decibels above threshold.

EQUAL TEMPERED SCALE

Frequency

C'	261.63	G	392.00
C#	277.18	G#	415.31
D	293.67	A	**440**
D#	311.13	A#	466.16
E	329.63	B	493.88
F	349.23	C"	523.25
F#	369.99		

Lowest frequency distinguishable as a note abt. 30 Hz
Highest audible frequency   abt. 40 000 Hz

GENERAL CONSTANTS

		Error (p.p.m.)
*Charge of the electron (e)	1.602192×10^{-19} C	4.4
	4.80325×10^{-10} e.s.u.	4.4
*Rest mass of the electron (m_e)		
	9.10956×10^{-31} kg	6.0
*Ratio of electron charge to rest mass (e/m_e)		
	1.758803×10^{11} C kg^{-1}	3.1
Velocity of light (c)	$299\,792\,458$ m s^{-1}	0.004
*Planck quantum of action (h)		
	6.62620×10^{-34} J s	7.6
	4.13571×10^{-15} eV s	10
$\hbar = h/2\pi$	1.05459×10^{-34} J s	7.6
*Boltzmann constant (k)	1.38062×10^{-23} J K^{-1}	43
*Gas constant (R_0)	8.31432 J K^{-1} mol^{-1}	42
No. of molecules per m³ at s.t.p.		
	2.68684×10^{25}	60
Volume of 1 mole of gas at s.t.p.		
	2.24136×10^{-2} m³	45
Mass of a hydrogen atom	1.67352×10^{-27} kg	15
Mass of molecule of molecular weight M		
	$M(1.66053) \times 10^{-27}$ kg	
Avogadro's constant (N)	6.0220943×10^{23} mol^{-1}	1
Quantum of wave number σ	$\sigma(1.98649) \times 10^{-25}$ J	8
Kinetic energy of a V volt electron		
	$V(1.60219) \times 10^{-19}$ J	
Wavelength of a V volt electron		
	$(V^{-\frac{1}{2}})(1.22643) \times 10^{-9}$ m	
*Gravitational constant (G)	6.6732×10^{-11} m³ kg^{-1} s^{-2}	500
*Stefan–Boltzmann constant (σ)		
	5.6696×10^{-8} W m^{-2} K^{-4}	200
*Proton/electron mass ratio	1836.109	6.2
*Magnetic flux quantum	2.067854×10^{-15} Wb	3.3
*Bohr radius (a_0)	5.29177×10^{-11} m	
*Bohr magneton	9.27410×10^{-24} J m² Wb^{-1}	7.0

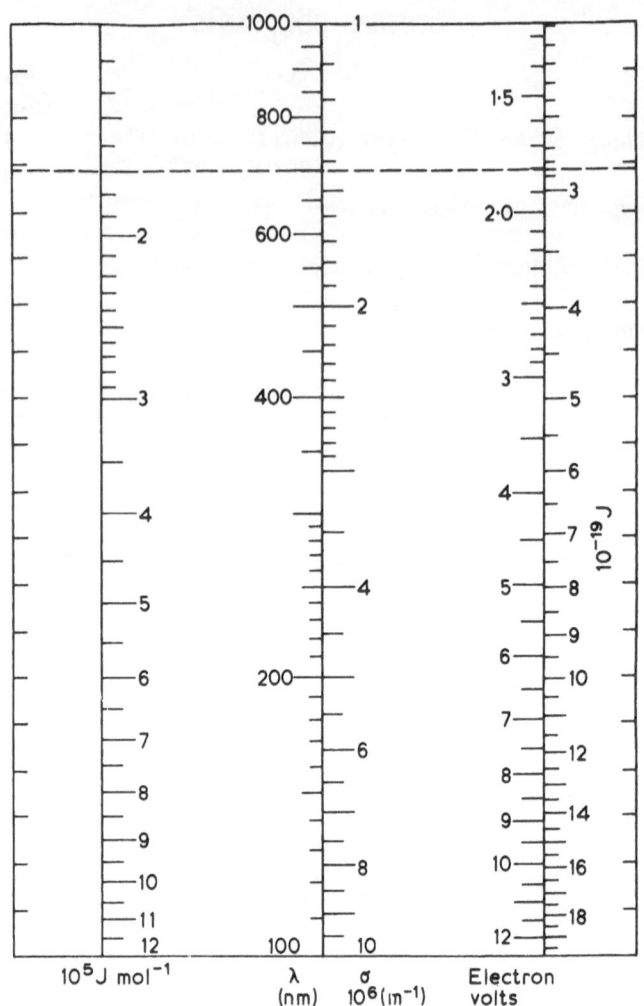

Nomogram 7. Quantum interconversion.

A straight edge placed *horizontally* across the diagram (use the two outer scales) permits rapid interconversion between energies expressed in (*a*) joule per mole (*b*) quanta of wavelength λ nm (*c*) quanta of wavenumber σ m^{-1} (*d*) electron volts (*e*) joule. For example, the equivalence is shown between quanta of wavelength 700 nm, wave-number 1.43×10^6 m^{-1}, and a 1.77 volt electron, which all have energies of 2.84×10^{-19} J, or when expressed per mole, of 1.71×10^5 J.

92

Nuclear magneton	5.05095×10^{-27} J m^2 Wb^{-1}	7.0
*Electron magnetic moment	9.28485×10^{-24} J m^2 Wb^{-1}	7.0
*Proton magnetic moment	1.410620×10^{-26} J m^2 Wb^{-1}	7.0
*Fine-structure constant (a)	7.29735×10^{-3} J m^{2}Wb^{-1}	1.5
$1/a$	137.0360	1.5

*Compton wavelength of the electron

2.42631×10^{-12} m $\qquad$ 3.1

*Classical electron radius $\quad 2.81794 \times 10^{-15}$ m $\qquad$ 4.6

Band spectrum constant $\quad 2.79933 \times 10^{-44}$ kg m $\qquad$ 8

Density of water (4°C, 1 atm) 999.9719 kg m^{-3}

Density of mercury (0°C, 1 atm)

$\qquad$ 13595.08 kg m^3

* Taylor, Parker and Langenberg values, see Preface.

Conversion factors

1 in $\quad = 2.54$ cm*	1 gr $\ = 0.064799$ g	1 in^3 $= 16.387$ cm^3
1 yd $\ = 0.9144$ m*	1 lb $\ = 0.45359237$ kg*	1 yd^3 $= 0.76455$ m^3
1 mile $= 1.609344$ km	1 ton $= 1016$ kg	1 gal $= 4.545963$ dm^3

1 fluid oz = 28.4131 cm^3; 1 oz (Troy or apothecaries') = 31.1035 g

* legal definitions.

Mathematical constants

$e = 2.7182818$	$\pi = 3.1415927$	$\log \pi = 0.4971499$
$e^{-1} = 0.3678794$	$\pi^2 = 9.8696044$	$\log \pi^2 = 0.9942997$
$\log_e 10 = 2.3025851$	$\pi^{-1} = 0.3183099$	$\sqrt{\pi} = 1.7724539$

1 radian = 57.29578°$\qquad$ 1° = 0.01745329 rad

LOGARITHMS

	0	1	2	3	4	5	6	7	8	9	Mean Differences.								
											1	2	3	4	5	6	7	8	9
10	0000	0043	0086	0128	0170	0212	0253	0294	0334	0374	4	8	12	17	21	25	29	33	37
11	0414	0453	0492	0531	0569	0607	0645	0682	0719	0755	4	8	11	15	19	23	26	30	34
12	0792	0828	0864	0899	0934	0969	1004	1038	1072	1106	3	7	10	14	17	21	24	28	31
13	1139	1173	1206	1239	1271	1303	1335	1367	1399	1430	3	6	10	13	16	19	23	26	29
14	1461	1492	1523	1553	1584	1614	1644	1673	1703	1732	3	6	9	12	15	18	21	24	27
15	1761	1790	1818	1847	1875	1903	1931	1959	1987	2014	3	6	8	11	14	17	20	22	25
16	2041	2068	2095	2122	2148	2175	2201	2227	2253	2279	3	5	8	11	13	16	18	21	24
17	2304	2330	2355	2380	2405	2430	2455	2480	2504	2529	2	5	7	10	12	15	17	20	22
18	2553	2577	2601	2625	2648	2672	2695	2718	2742	2765	2	5	7	9	12	14	16	19	21
19	2788	2810	2833	2856	2878	2900	2923	2945	2967	2989	2	4	7	9	11	13	16	18	20
20	3010	3032	3054	3075	3096	3118	3139	3160	3181	3201	2	4	6	8	11	13	15	17	19
21	3222	3243	3263	3284	3304	3324	3345	3365	3385	3404	2	4	6	8	10	12	14	16	18
22	3424	3444	3464	3483	3502	3522	3541	3560	3579	3598	2	4	6	8	10	12	14	15	17
23	3617	3636	3655	3674	3692	3711	3729	3747	3766	3784	2	4	6	7	9	11	13	15	17
24	3802	3820	3838	3856	3874	3892	3909	3927	3945	3962	2	4	5	7	9	11	12	14	16
25	3979	3997	4014	4031	4048	4065	4082	4099	4116	4133	2	3	5	7	9	10	12	14	15
26	4150	4166	4183	4200	4216	4232	4249	4265	4281	4298	2	3	5	7	8	10	11	13	15
27	4314	4330	4346	4362	4378	4393	4409	4425	4440	4456	2	3	5	6	8	9	11	13	14
28	4472	4487	4502	4518	4533	4548	4564	4579	4594	4609	2	3	5	6	8	9	11	12	14
29	4624	4639	4654	4669	4683	4698	4713	4728	4742	4757	1	3	4	6	7	9	10	12	13
30	4771	4786	4800	4814	4829	4843	4857	4871	4886	4900	1	3	4	6	7	9	10	11	13
31	4914	4928	4942	4955	4969	4983	4997	5011	5024	5038	1	3	4	6	7	8	10	11	12
32	5051	5065	5079	5092	5105	5119	5132	5145	5159	5172	1	3	4	5	7	8	9	11	12
33	5185	5198	5211	5224	5237	5250	5263	5276	5289	5302	1	3	4	5	6	8	9	10	12
34	5315	5328	5340	5353	5366	5378	5391	5403	5416	5428	1	3	4	5	6	8	9	10	11
35	5441	5453	5465	5478	5490	5502	5514	5527	5539	5551	1	2	4	5	6	7	9	10	11
36	5563	5575	5587	5599	5611	5623	5635	5647	5658	5670	1	2	4	5	6	7	8	10	11
37	5682	5694	5705	5717	5729	5740	5752	5763	5775	5786	1	2	3	5	6	7	8	9	10
38	5798	5809	5821	5832	5843	5855	5866	5877	5888	5899	1	2	3	5	6	7	8	9	10
39	5911	5922	5933	5944	5955	5966	5977	5988	5999	6010	1	2	3	4	5	7	8	9	10
40	6021	6031	6042	6053	6064	6075	6085	6096	6107	6117	1	2	3	4	5	6	8	9	10
41	6128	6138	6149	6160	6170	6180	6191	6201	6212	6222	1	2	3	4	5	6	7	8	9
42	6232	6243	6253	6263	6274	6284	6294	6304	6314	6325	1	2	3	4	5	6	7	8	9
43	6335	6345	6355	6365	6375	6385	6395	6405	6415	6425	1	2	3	4	5	6	7	8	9
44	6435	6444	6454	6464	6474	6484	6493	6503	6513	6522	1	2	3	4	5	6	7	8	9
45	6532	6542	6551	6561	6571	6580	6590	6599	6609	6618	1	2	3	4	5	6	7	8	9
46	6628	6637	6646	6656	6665	6675	6684	6693	6702	6712	1	2	3	4	5	6	7	7	8
47	6721	6730	6739	6749	6758	6767	6776	6785	6794	6803	1	2	3	4	5	5	6	7	8
48	6812	6821	6830	6839	6848	6857	6866	6875	6884	6893	1	2	3	4	4	5	6	7	8
49	6902	6911	6920	6928	6937	6946	6955	6964	6972	6981	1	2	3	4	4	5	6	7	8
50	6990	6998	7007	7016	7024	7033	7042	7050	7059	7067	1	2	3	3	4	5	6	7	8
51	7076	7084	7093	7101	7110	7118	7126	7135	7143	7152	1	2	3	3	4	5	6	7	8
52	7160	7168	7177	7185	7193	7202	7210	7218	7226	7235	1	2	2	3	4	5	6	7	7
53	7243	7251	7259	7267	7275	7284	7292	7300	7308	7316	1	2	2	3	4	5	6	6	7
54	7324	7332	7340	7348	7356	7364	7372	7380	7388	7396	1	2	2	3	4	5	6	6	7

LOGARITHMS

	0	1	2	3	4	5	6	7	8	9	1	2	3	4	5	6	7	8	9
														Mean Differences.					
55	7404	7412	7419	7427	7435	**7443**	7451	7459	7466	7474	1	2	2	3	4	5	5	6	7
56	7482	7490	7497	7505	7513	**7520**	7528	7536	7543	7551	1	2	2	3	4	5	5	6	7
57	7559	7566	7574	7582	7589	**7597**	7604	7612	7619	7627	1	2	2	3	4	5	5	6	7
58	7634	7642	7649	7657	7664	**7672**	7679	7686	7694	7701	1	1	2	3	4	4	5	6	7
59	7709	7716	7723	7731	7738	**7745**	7752	7760	7767	7774	1	1	2	3	4	4	5	6	7
60	7782	7789	7796	7803	7810	**7818**	7825	7832	7839	7846	1	1	2	3	4	4	5	6	6
61	7853	7860	7868	7875	7882	**7889**	7896	7903	7910	7917	1	1	2	3	4	4	5	6	6
62	7924	7931	7938	7945	7952	**7959**	7966	7973	7980	7987	1	1	2	3	3	4	5	6	6
63	7993	8000	8007	8014	8021	**8028**	8035	8041	8048	8055	1	1	2	3	3	4	5	5	6
64	8062	8069	8075	8082	8089	**8096**	8102	8109	8116	8122	1	1	2	3	3	4	5	5	6
65	8129	8136	8142	8149	8156	**8162**	8169	8176	8182	8189	1	1	2	3	3	4	5	5	6
66	8195	8202	8209	8215	8222	**8228**	8235	8241	8248	8254	1	1	2	3	3	4	5	5	6
67	8261	8267	8274	8280	8287	**8293**	8299	8306	8312	8319	1	1	2	3	3	4	5	5	6
68	8325	8331	8338	8344	8351	**8357**	8363	8370	8376	8382	1	1	2	3	3	4	4	5	6
69	8388	8395	8401	8407	8414	**8420**	8426	8432	8439	8445	1	1	2	2	3	4	4	5	6
70	8451	8457	8463	8470	8476	**8482**	8488	8494	8500	8506	1	1	2	2	3	4	4	5	6
71	8513	8519	8525	8531	8537	**8543**	8549	8555	8561	8567	1	1	2	2	3	4	4	5	5
72	8573	8579	8585	8591	8597	**8603**	8609	8615	8621	8627	1	1	2	2	3	4	4	5	5
73	8633	8639	8645	8651	8657	**8663**	8669	8675	8681	8686	1	1	2	2	3	4	4	5	5
74	8692	8698	8704	8710	8716	**8722**	8727	8733	8739	8745	1	1	2	2	3	4	4	5	5
75	8751	8753	8762	8768	8774	**8779**	8785	8791	8797	8802	1	1	2	2	3	3	4	5	5
76	8808	8814	8820	8825	8831	**8837**	8842	8848	8854	8859	1	1	2	2	3	3	4	5	5
77	8865	8871	8876	8882	8887	**8893**	8899	8904	8910	8915	1	1	2	2	3	3	4	4	5
78	8921	8927	8932	8938	8943	**8949**	8954	8960	8965	8971	1	1	2	2	3	3	4	4	5
79	8976	8982	8987	8993	8998	**9004**	9009	9015	9020	9025	1	1	2	2	3	3	4	4	5
80	9031	9036	9042	9047	9053	**9058**	9063	9069	9074	9079	1	1	2	2	3	3	4	4	5
81	9085	9090	9096	9101	9106	**9112**	9117	9122	9128	9133	1	1	2	2	3	3	4	4	5
82	9138	9143	9149	9154	9159	**9165**	9170	9175	9180	9186	1	1	2	2	3	3	4	4	5
83	9191	9196	9201	9206	9212	**9217**	9222	9227	9232	9238	1	1	2	2	3	3	4	4	5
84	9243	9248	9253	9258	9263	**9269**	9274	9279	9284	9289	1	1	2	2	3	3	4	4	5
85	9294	9299	9304	9309	9315	**9320**	9325	9330	9335	9340	1	1	2	2	3	3	4	4	5
86	9345	9350	9355	9360	9365	**9370**	9375	9380	9385	9390	1	1	2	2	3	3	4	4	5
87	9395	9400	9405	9410	9415	**9420**	9425	9430	9435	9440	0	1	1	2	2	3	3	4	4
88	9445	9450	9455	9460	9465	**9469**	9474	9479	9484	9489	0	1	1	2	2	3	3	4	4
89	9494	9499	9504	9509	9513	**9518**	9523	9528	9533	9538	0	1	1	2	2	3	3	4	4
90	9542	9547	9552	9557	9562	**9566**	9571	9576	9581	9586	0	1	1	2	2	3	3	4	4
91	9590	9595	9600	9605	9609	**9614**	9619	9624	9628	9633	0	1	1	2	2	3	3	4	4
92	9638	9643	9647	9652	9657	**9661**	9666	9671	9675	9680	0	1	1	2	2	3	3	4	4
93	9685	9689	9694	9699	9703	**9708**	9713	9717	9722	9727	0	1	1	2	2	3	3	4	4
94	9731	9736	9741	9745	9750	**9754**	9759	9763	9768	9773	0	1	1	2	2	3	3	4	4
95	9777	9782	9786	9791	9795	**9800**	9805	9809	9814	9818	0	1	1	2	2	3	3	4	4
96	9823	9827	9832	9836	9841	**9845**	9850	9854	9859	9863	0	1	1	2	2	3	3	4	4
97	9868	9872	9877	9881	9886	**9890**	9894	9899	9903	9908	0	1	1	2	2	3	3	4	4
98	9912	9917	9921	9926	9930	**9934**	9939	9943	9948	9952	0	1	1	2	2	3	3	4	4
99	9956	9961	9965	9969	9974	**9978**	9983	9987	9991	9996	0	1	1	2	2	3	3	3	4

ANTILOGARITHMS

	0	1	2	3	4	5	6	7	8	9	Mean Differences.								
											1	2	3	4	5	6	7	8	9
·00	1000	1002	1005	1007	1009	**1012**	1014	1016	1019	1021	0	1	1	1	1	2	2	2	2
·01	1023	1026	1028	1030	1033	**1035**	1038	1040	1042	1045	0	1	1	1	1	2	2	2	2
·02	1047	1050	1052	1054	1057	**1059**	1062	1064	1067	1069	0	1	1	1	1	2	2	2	2
·03	1072	1074	1076	1079	1081	**1084**	1086	1089	1091	1094	0	1	1	1	1	2	2	2	2
·04	1096	1099	1102	1104	1107	**1109**	1112	1114	1117	1119	1	1	1	1	1	2	2	2	2
·05	1122	1125	1127	1130	1132	**1135**	1138	1140	1143	1146	0	1	1	1	1	2	2	2	2
·06	1148	1151	1153	1156	1159	**1161**	1164	1167	1169	1172	0	1	1	1	1	2	2	2	2
·07	1175	1178	1180	1183	1186	**1189**	1191	1194	1197	1199	0	1	1	1	1	2	2	2	2
·08	1202	1205	1208	1211	1213	**1216**	1219	1222	1225	1227	0	1	1	1	1	2	2	2	3
·09	1230	1233	1236	1239	1242	**1245**	1247	1250	1253	1256	0	1	1	1	1	2	2	2	3
·10	1259	1262	1265	1268	1271	**1274**	1276	1279	1282	1285	0	1	1	1	1	2	2	2	3
·11	1288	1291	1294	1297	1300	**1303**	1306	1309	1312	1315	0	1	1	1	2	2	2	2	3
·12	1318	1321	1324	1327	1330	**1334**	1337	1340	1343	1346	0	1	1	1	2	2	2	2	3
·13	1349	1352	1355	1358	1361	**1365**	1368	1371	1374	1377	0	1	1	1	2	2	2	3	3
·14	1380	1384	1387	1390	1393	**1396**	1400	1403	1406	1409	0	1	1	1	2	2	2	3	3
·15	1413	1416	1419	1422	1426	**1429**	1432	1435	1439	1442	0	1	1	1	2	2	2	3	3
·16	1445	1449	1452	1455	1459	**1462**	1466	1469	1472	1476	0	1	1	1	2	2	2	3	3
·17	1479	1483	1486	1489	1493	**1496**	1500	1503	1507	1510	0	1	1	1	2	2	2	3	3
·18	1514	1517	1521	1524	1528	**1531**	1535	1538	1542	1545	0	1	1	1	2	2	2	3	3
·19	1549	1552	1556	1560	1563	**1567**	1570	1574	1578	1581	0	1	1	1	2	2	3	3	3
·20	1585	1589	1592	1596	1600	**1603**	1607	1611	1614	1618	0	1	1	1	2	2	3	3	3
·21	1622	1626	1629	1633	1637	**1641**	1644	1648	1652	1656	0	1	1	2	2	2	3	3	3
·22	1660	1663	1667	1671	1675	**1679**	1683	1687	1690	1694	0	1	1	2	2	2	3	3	3
·23	1698	1702	1706	1710	1714	**1718**	1722	1726	1730	1734	0	1	1	2	2	2	3	3	4
·24	1738	1742	1746	1750	1754	**1758**	1762	1766	1770	1774	0	1	1	2	2	2	3	3	4
·25	1778	1782	1786	1791	1795	**1799**	1803	1807	1811	1816	0	1	1	2	2	2	3	3	4
·26	1820	1824	1828	1832	1837	**1841**	1845	1849	1854	1858	0	1	1	2	2	3	3	3	4
·27	1862	1866	1871	1875	1879	**1884**	1888	1892	1897	1901	0	1	1	2	2	3	3	3	4
·28	1905	1910	1914	1919	1923	**1928**	1932	1936	1941	1945	0	1	1	2	2	3	3	4	4
·29	1950	1954	1959	1963	1968	**1972**	1977	1982	1986	1991	0	1	1	2	2	3	3	4	4
·30	1995	2000	2004	2009	2014	**2018**	2023	2028	2032	2037	0	1	1	2	2	3	3	4	4
·31	2042	2046	2051	2056	2061	**2065**	2070	2075	2080	2084	0	1	1	2	2	3	3	4	4
·32	2089	2094	2099	2104	2109	**2113**	2118	2123	2128	2133	0	1	1	2	2	3	3	4	4
·33	2138	2143	2148	2153	2158	**2163**	2168	2173	2178	2183	0	1	1	2	2	3	3	4	4
·34	2188	2193	2198	2203	2208	**2213**	2218	2223	2228	2234	1	1	2	2	3	3	4	4	5
·35	2239	2244	2249	2254	2259	**2265**	2270	2275	2280	2286	1	1	2	2	3	3	4	4	5
·36	2291	2296	2301	2307	2312	**2317**	2323	2328	2333	2339	1	1	2	2	3	3	4	4	5
·37	2344	2350	2355	2360	2366	**2371**	2377	2382	2388	2393	1	1	2	2	3	3	4	4	5
·38	2399	2404	2410	2415	2421	**2427**	2432	2438	2443	2449	1	1	2	2	3	3	4	4	5
·39	2455	2460	2466	2472	2477	**2483**	2489	2495	2500	2506	1	1	2	2	3	3	4	5	5
·40	2512	2518	2523	2529	2535	**2541**	2547	2553	2559	2564	1	1	2	2	3	4	4	5	5
·41	2570	2576	2582	2588	2594	**2600**	2606	2612	2618	2624	1	1	2	2	3	4	4	5	5
·42	2630	2636	2642	2649	2655	**2661**	2667	2673	2679	2685	1	1	2	2	3	4	4	5	6
·43	2692	2698	2704	2710	2716	**2723**	2729	2735	2742	2748	1	1	2	3	3	4	4	5	6
·44	2754	2761	2767	2773	2780	**2786**	2793	2799	2805	2812	1	1	2	3	3	4	4	5	6
·45	2818	2825	2831	2838	2844	**2851**	2858	2864	2871	2877	1	1	2	3	3	4	5	5	6
·46	2884	2891	2897	2904	2911	**2917**	2924	2931	2938	2944	1	1	2	3	4	4	5	5	6
·47	2951	2958	2965	2972	2979	**2985**	2992	2999	3006	3013	1	1	2	3	3	4	5	5	6
·48	3020	3027	3034	3041	3048	**3055**	3062	3069	3076	3083	1	1	2	3	4	4	5	6	6
·49	3090	3097	3105	3112	3119	**3126**	3133	3141	3148	3155	1	1	2	3	4	4	5	6	6

ANTILOGARITHMS

	0	1	2	3	4	5	6	7	8	9	|1	2	3	4	5	6	7	8	9
													Mean	Differences.					
·50	3162	3170	3177	3184	3192	**3199**	3206	3214	3221	3228	1	1	2	3	4	4	5	6	7
·51	3236	3243	3251	3258	3266	**3273**	3281	3289	3296	3304	1	2	2	3	4	5	5	6	7
·52	3311	3319	3327	3334	3342	**3350**	3357	3365	3373	3381	1	2	2	3	4	5	5	6	7
·53	3388	3396	3404	3412	3420	**3428**	3436	3443	3451	3459	1	2	2	3	4	5	6	6	7
·54	3467	3475	3483	3491	3499	**3508**	3516	3524	3532	3540	1	2	2	3	4	5	6	6	7
·55	3548	3556	3565	3573	3581	**3589**	3597	3606	3614	3622	1	2	2	3	4	5	6	7	7
·56	3631	3639	3648	3656	3664	**3673**	3681	3690	3698	3707	1	2	3	3	4	5	6	7	8
·57	3715	3724	3733	3741	3750	**3758**	3767	3776	3784	3793	1	2	3	3	4	5	6	7	8
·58	3802	3811	3819	3828	3837	**3846**	3855	3864	3873	3882	1	2	3	4	4	5	6	7	8
·59	3890	3899	3908	3917	3926	**3936**	3945	3954	3963	3972	1	2	3	4	5	5	6	7	8
·60	3981	3990	3999	4009	4018	**4027**	4036	4046	4055	4064	1	2	3	4	5	6	6	7	8
·61	4074	4083	4093	4102	4111	**4121**	4130	4140	4150	4159	1	2	3	4	5	6	7	8	9
·62	4169	4178	4188	4198	4207	**4217**	4227	4236	4246	4256	1	2	3	4	5	6	7	8	9
·63	4266	4276	4285	4295	4305	**4315**	4325	4335	4345	4355	1	2	3	4	5	6	7	8	9
·64	4365	4375	4385	4395	4406	**4416**	4426	4436	4446	4457	1	2	3	4	5	6	7	8	9
·65	4467	4477	4487	4498	4508	**4519**	4529	4539	4550	4560	1	2	3	4	5	6	7	8	9
·66	4571	4581	4592	4603	4613	**4624**	4634	4645	4656	4667	1	2	3	4	5	6	7	9	10
·67	4677	4688	4699	4710	4721	**4732**	4742	4753	4764	4775	1	2	3	4	5	7	8	9	10
·68	4786	4797	4808	4819	4831	**4842**	4853	4864	4875	4887	1	2	3	4	6	7	8	9	10
·69	4898	4909	4920	4932	4943	**4955**	4966	4977	4989	5000	1	2	3	5	6	7	8	9	10
·70	5012	5023	5035	5047	5058	**5070**	5082	5093	5105	5117	1	2	4	5	6	7	8	9	11
·71	5129	5140	5152	5164	5176	**5188**	5200	5212	5224	5236	1	2	4	5	6	7	8	10	11
·72	5248	5260	5272	5284	5297	**5309**	5321	5333	5346	5358	1	2	4	5	6	7	9	10	11
·73	5370	5383	5395	5408	5420	**5433**	5445	5458	5470	5483	1	3	4	5	6	8	9	10	11
·74	5495	5508	5521	5534	5546	**5559**	5572	5585	5598	5610	1	3	4	5	6	8	9	10	12
·75	5623	5636	5649	5662	5675	**5689**	5702	5715	5728	5741	1	3	4	5	7	8	9	10	12
·76	5754	5768	5781	5794	5808	**5821**	5834	5848	5861	5875	1	3	4	5	7	8	9	11	12
·77	5888	5902	5916	5929	5943	**5957**	5970	5984	5998	6012	1	3	4	5	7	8	10	11	12
·78	6026	6039	6053	6067	6081	**6095**	6109	6124	6138	6152	1	3	4	6	7	8	10	11	13
·79	6166	6180	6194	6209	6223	**6237**	6252	6266	6281	6295	1	3	4	6	7	9	10	11	13
·80	6310	6324	6339	6353	6368	**6383**	6397	6412	6427	6442	1	3	4	6	7	9	10	12	13
·81	6457	6471	6486	6501	6516	**6531**	6546	6561	6577	6592	2	3	5	6	8	9	11	12	14
·82	6607	6622	6637	6653	6668	**6683**	6699	6714	6730	6745	2	3	5	6	8	9	11	12	14
·83	6761	6776	6792	6808	6823	**6839**	6855	6871	6887	6902	2	3	5	6	8	9	11	13	14
·84	6918	6934	6950	6966	6982	**6998**	7015	7031	7047	7063	2	3	5	6	8	10	11	13	15
·85	7079	7096	7112	7129	7145	**7161**	7178	7194	7211	7228	2	3	5	7	8	10	12	13	15
·86	7244	7261	7278	7295	7311	**7328**	7345	7362	7379	7396	2	3	5	7	8	10	12	13	15
·87	7413	7430	7447	7464	7482	**7499**	7516	7534	7551	7568	2	3	5	7	9	10	12	14	16
·88	7586	7603	7621	7638	7656	**7674**	7691	7709	7727	7745	2	4	5	7	9	11	12	14	16
·89	7762	7780	7798	7816	7834	**7852**	7870	7889	7907	7925	2	4	5	7	9	11	13	14	16
·90	7943	7962	7980	7998	8017	**8035**	8054	8072	8091	8110	2	4	6	7	9	11	13	15	17
·91	8128	8147	8166	8185	8204	**8222**	8241	8260	8279	8299	2	4	6	8	9	11	13	15	17
·92	8318	8337	8356	8375	8395	**8414**	8433	8453	8472	8492	2	4	6	8	10	12	14	15	17
·93	8511	8531	8551	8570	8590	**8610**	8630	8650	8670	8690	2	4	6	8	10	12	14	16	18
·94	8710	8730	8750	8770	8790	**8810**	8831	8851	8872	8892	2	4	6	8	10	12	14	16	18
·95	8913	8933	8954	8974	8995	**9016**	9036	9057	9078	9099	2	4	6	8	10	12	15	17	19
·96	9120	9141	9162	9183	9204	**9226**	9247	9268	9290	9311	2	4	6	8	11	13	15	17	19
·97	9333	9354	9376	9397	9419	**9441**	9462	9484	9506	9528	2	4	7	9	11	13	15	17	20
·98	9550	9572	9594	9616	9638	**9661**	9683	9705	9727	9750	2	4	7	9	11	13	16	18	20
·99	9772	9795	9817	9840	9863	**9886**	9908	9931	9954	9977	2	5	7	9	11	14	16	18	20

NATURAL SINES

	0′	6′	12′	18′	24′	30′	36′	42′	48′	54′	Mean Differences.				
											1′	2′	3′	4′	5′
0°	·0000	0017	0035	0052	0070	**0087**	0105	0122	0140	0157	3	6	9	12	15
1	·0175	0192	0209	0227	0244	**0262**	0279	0297	0314	0332	3	6	9	12	15
2	·0349	0366	0384	0401	0419	**0436**	0454	0471	0488	0506	3	6	9	12	15
3	·0523	0541	0558	0576	0593	**0610**	0628	0645	0663	0680	3	6	9	12	15
4	·0698	0715	0732	0750	0767	**0785**	0802	0819	0837	0854	3	6	9	12	14
5	·0872	0889	0906	0924	0941	**0958**	0976	0993	1011	1028	3	6	9	12	14
6	·1045	1063	1080	1097	1115	**1132**	1149	1167	1184	1201	3	6	9	12	14
7	·1219	1236	1253	1271	1288	**1305**	1323	1340	1357	1374	3	6	9	12	14
8	·1392	1409	1426	1444	1461	**1478**	1495	1513	1530	1547	3	6	9	12	14
9	·1564	1582	1599	1616	1633	**1650**	1668	1685	1702	1719	3	6	9	12	14
10°	·1736	1754	1771	1788	1805	**1822**	1840	1857	1874	1891	3	6	9	11	14
11	·1908	1925	1942	1959	1977	**1994**	2011	2028	2045	2062	3	6	9	11	14
12	·2079	2096	2113	2130	2147	**2164**	2181	2198	2215	2233	3	6	9	11	14
13	·2250	2267	2284	2300	2317	**2334**	2351	2368	2385	2402	3	6	8	11	14
14	·2419	2436	2453	2470	2487	**2504**	2521	2538	2554	2571	3	6	8	11	14
15	·2588	2605	2622	2639	2656	**2672**	2689	2706	2723	2740	3	6	8	11	14
16	·2756	2773	2790	2807	2823	**2840**	2857	2874	2890	2907	3	6	8	11	14
17	·2924	2940	2957	2974	2990	**3007**	3024	3040	3057	3074	3	6	8	11	14
18	·3090	3107	3123	3140	3156	**3173**	3190	3206	3223	3239	3	6	8	11	14
19	·3256	3272	3289	3305	3322	**3338**	3355	3371	3387	3404	3	5	8	11	14
20°	·3420	3437	3453	3469	3486	**3502**	3518	3535	3551	3567	3	5	8	11	14
21	·3584	3600	3616	3633	3649	**3665**	3681	3697	3714	3730	3	5	8	11	14
22	·3746	3762	3778	3795	3811	**3827**	3843	3859	3875	3891	3	5	8	11	14
23	·3907	3923	3939	3955	3971	**3987**	4003	4019	4035	4051	3	5	8	11	14
24	·4067	4083	4099	4115	4131	**4147**	4163	4179	4195	4210	3	5	8	11	13
25	·4226	4242	4258	4274	4289	**4305**	4321	4337	4352	4368	3	5	8	11	13
26	·4384	4399	4415	4431	4446	**4462**	4478	4493	4509	4524	3	5	8	10	13
27	·4540	4555	4571	4586	4602	**4617**	4633	4648	4664	4679	3	5	8	10	13
28	·4695	4710	4726	4741	4756	**4772**	4787	4802	4818	4833	3	5	8	10	13
29	·4848	4863	4879	4894	4909	**4924**	4939	4955	4970	4985	3	5	8	10	13
30°	·5000	5015	5030	5045	5060	**5075**	5090	5105	5120	5135	3	5	8	10	13
31	·5150	5165	5180	5195	5210	**5225**	5240	5255	5270	5284	2	5	7	10	12
32	·5299	5314	5329	5344	5358	**5373**	5388	5402	5417	5432	2	5	7	10	12
33	·5446	5461	5476	5490	5505	**5519**	5534	5548	5563	5577	2	5	7	10	12
34	·5592	5606	5621	5635	5650	**5664**	5678	5693	5707	5721	2	5	7	10	12
35	·5736	5750	5764	5779	5793	**5807**	5821	5835	5850	5864	2	5	7	9	12
36	·5878	5892	5906	5920	5934	**5948**	5962	5976	5990	6004	2	5	7	9	12
37	·6018	6032	6046	6060	6074	**6088**	6101	6115	6129	6143	2	5	7	9	12
38	·6157	6170	6184	6198	6211	**6225**	6239	6252	6266	6280	2	5	7	9	11
39	·6293	6307	6320	6334	6347	**6361**	6374	6388	6401	6414	2	4	7	9	11
40°	·6428	6441	6455	6468	6481	**6494**	6508	6521	6534	6547	2	4	7	9	11
41	·6561	6574	6587	6600	6613	**6626**	6639	6652	6665	6678	2	4	7	9	11
42	·6691	6704	6717	6730	6743	**6756**	6769	6782	6794	6807	2	4	6	9	11
43	·6820	6833	6845	6858	6871	**6884**	6896	6909	6921	6934	2	4	6	8	11
44	·6947	6959	6972	6984	6997	**7009**	7022	7034	7046	7059	2	4	6	8	10

98

NATURAL SINES

	0′	6′	12′	18′	24′	30′	36′	42′	48′	54′	Mean Differences.				
											1′	2′	3′	4′	5′
45°	·7071	7083	7096	7108	7120	**7133**	7145	7157	7169	7181	2	4	6	8	10
46	·7193	7206	7218	7230	7242	**7254**	7266	7278	7290	7302	2	4	6	8	10
47	·7314	7325	7337	7349	7361	**7373**	7385	7396	7408	7420	2	4	6	8	10
48	·7431	7443	7455	7466	7478	**7490**	7501	7513	7524	7536	2	4	6	8	10
49	·7547	7559	7570	7581	7593	**7604**	7615	7627	7638	7649	2	4	6	8	9
50°	·7660	7672	7683	7694	7705	**7716**	7727	7738	7749	7760	2	4	6	7	9
51	·7771	7782	7793	7804	7815	**7826**	7837	7848	7859	7869	2	4	5	7	9
52	·7880	7891	7902	7912	7923	**7934**	7944	7955	7965	7976	2	4	5	7	9
53	·7986	7997	8007	8018	8028	**8039**	8049	8059	8070	8080	2	3	5	7	9
54	·8090	8100	8111	8121	8131	**8141**	8151	8161	8171	8181	2	3	5	7	8
55	·8192	8202	8211	8221	8231	**8241**	8251	8261	8271	8281	2	3	5	7	8
56	·8290	8300	8310	8320	8329	**8339**	8348	8358	8368	8377	2	3	5	6	8
57	·8387	8396	8406	8415	8425	**8434**	8443	8453	8462	8471	2	3	5	6	8
58	·8480	8490	8499	8508	8517	**8526**	8536	8545	8554	8563	2	3	5	6	8
59	·8572	8581	8590	8599	8607	**8616**	8625	8634	8643	8652	1	3	4	6	7
60°	·8660	8669	8678	8686	8695	**8704**	8712	8721	8729	8738	1	3	4	6	7
61	·8746	8755	8763	8771	8780	**8788**	8796	8805	8813	8821	1	3	4	6	7
62	·8829	8838	8846	8854	8862	**8870**	8878	8886	8894	8902	1	3	4	5	7
63	·8910	8918	8926	8934	8942	**8949**	8957	8965	8973	8980	1	3	4	5	6
64	·8988	8996	9003	9011	9018	**9026**	9033	9041	9048	9056	1	3	4	5	6
65	·9063	9070	9078	9085	9092	**9100**	9107	9114	9121	9128	1	2	4	5	6
66	·9135	9143	9150	9157	9164	**9171**	9178	9184	9191	9198	1	2	3	5	6
67	·9205	9212	9219	9225	9232	**9239**	9245	9252	9259	9265	1	2	3	4	6
68	·9272	9278	9285	9291	9298	**9304**	9311	9317	9323	9330	1	2	3	4	5
69	·9336	9342	9348	9354	9361	**9367**	9373	9379	9385	9391	1	2	3	4	5
70°	·9397	9403	9409	9415	9421	**9426**	9432	9438	9444	9449	1	2	3	4	5
71	·9455	9461	9466	9472	9478	**9483**	9489	9494	9500	9505	1	2	3	4	5
72	·9511	9516	9521	9527	9532	**9537**	9542	9548	9553	9558	1	2	3	3	4
73	·9563	9568	9573	9578	9583	**9588**	9593	9598	9603	9608	1	2	2	3	4
74	·9613	9617	9622	9627	9632	**9636**	9641	9646	9650	9655	1	2	2	3	4
75	·9659	9664	9668	9673	9677	**9681**	9686	9690	9694	9699	1	1	2	3	4
76	·9703	9707	9711	9715	9720	**9724**	9728	9732	9736	9740	1	1	2	3	3
77	·9744	9748	9751	9755	9759	**9763**	9767	9770	9774	9778	1	1	2	3	3
78	·9781	9785	9789	9792	9796	**9799**	9803	9806	9810	9813	1	1	2	2	3
79	·9816	9820	9823	9826	9829	**9833**	9836	9839	9842	9845	1	1	2	2	3
80°	·9848	9851	9854	9857	9860	**9863**	9866	9869	9871	9874	0	1	1	2	2
81	·9877	9880	9882	9885	9888	**9890**	9893	9895	9898	9900	0	1	1	2	2
82	·9903	9905	9907	9910	9912	**9914**	9917	9919	9921	9923	0	1	1	2	2
83	·9925	9928	9930	9932	9934	**9936**	9938	9940	9942	9943	0	1	1	1	2
84	·9945	9947	9949	9951	9952	**9954**	9956	9957	9959	9960	0	1	1	1	2
85	·9962	9963	9965	9966	9968	**9969**	9971	9972	9973	9974	**0**	**0**	1	1	1
86	·9976	9977	9978	9979	9980	**9981**	9982	9983	9984	9985	0	0	1	1	1
87	·9986	9987	9988	9989	9990	**9990**	9991	9992	9993	9993	0	0	0	1	1
88	·9994	9995	9995	9996	9996	**9997**	9997	9997	9998	9998	0	0	0	0	0
89	·9998	9999	9999	9999	9999	**1·000**	1·000	1·000	1·000	1·000	0	0	0	0	0

NATURAL COSINES

(N.B., Subtract Mean Differences)

	0′	6′	12′	18′	24′	30′	36′	42′	48′	54′	Mean Differences. 1′	2′	3′	4′	5′
0°	1·000	1·000	1·000	1·000	1·000	1·000	·9999	9999	9999	9999	0	0	0	0	0
1	·9998	9998	9998	9997	9997	9997	9996	9996	9995	9995	0	0	0	0	0
2	·9994	9993	9993	9992	9991	9990	9990	9989	9988	9987	0	0	0	1	1
3	·9986	9985	9984	9983	9982	9981	9980	9979	9978	9977	0	0	1	1	1
4	·9976	9974	9973	9972	9971	9969	9968	9966	9965	9963	0	0	1	1	1
5	·9962	9960	9959	9957	9956	9954	9952	9951	9949	9947	0	1	1	1	2
6	·9945	9943	9942	9940	9938	9936	9934	9932	9930	9928	0	1	1	1	2
7	·9925	9923	9921	9919	9917	9914	9912	9910	9907	9905	0	1	1	2	2
8	·9903	9900	9898	9895	9893	9890	9888	9885	9882	9880	0	1	1	2	2
9	·9877	9874	9871	9869	9866	9863	9860	9857	9854	9851	0	1	1	2	2
10°	·9848	9845	9842	9839	9836	9833	9829	9826	9823	9820	1	1	2	2	3
11	·9816	9813	9810	9806	9803	9799	9796	9792	9789	9785	1	1	2	2	3
12	·9781	9778	9774	9770	9767	9763	9759	9755	9751	9748	1	1	2	3	3
13	·9744	9740	9736	9732	9728	9724	9720	9715	9711	9707	1	1	2	3	3
14	·9703	9699	9694	9690	9686	9681	9677	9673	9668	9664	1	1	2	3	4
15	·9659	9655	9650	9646	9641	9636	9632	9627	9622	9617	1	2	2	3	4
16	·9613	9608	9603	9598	9593	9588	9583	9578	9573	9568	1	2	2	3	4
17	·9553	9558	9553	9548	9542	9537	9532	9527	9521	9516	1	2	3	3	4
18	·9511	9505	9500	9494	9489	9483	9478	9472	9466	9461	1	2	3	4	5
19	·9455	9449	9444	9438	9432	9426	9421	9415	9409	9403	1	2	3	4	5
20°	·9397	9391	9385	9379	9373	9367	9361	9354	9348	9342	1	2	3	4	5
21	·9336	9330	9323	9317	9311	9304	9298	9291	9285	9278	1	2	3	4	5
22	·9272	9265	9259	9252	9245	9239	9232	9225	9219	9212	1	2	3	4	6
23	·9205	9198	9191	9184	9178	9171	9164	9157	9150	9143	1	2	3	5	6
24	·9135	9128	9121	9114	9107	9100	9092	9085	9078	9070	1	2	4	5	6
25	·9063	9056	9048	9041	9033	9026	9018	9011	9003	8996	1	3	4	5	6
26	·8988	8980	8973	8965	8957	8949	8942	8934	8926	8918	1	3	4	5	6
27	·8910	8902	8894	8886	8878	8870	8862	8854	8846	8838	1	3	4	5	7
28	·8829	8821	8813	8805	8796	8788	8780	8771	8763	8755	1	3	4	6	7
29	·8746	8738	8729	8721	8712	8704	8695	8686	8678	8669	1	3	4	6	7
30°	·8660	8652	8643	8634	8625	8616	8607	8599	8590	8581	1	3	4	6	7
31	·8572	8563	8554	8545	8536	8526	8517	8508	8499	8490	2	3	5	6	8
32	·8480	8471	8462	8453	8443	8434	8425	8415	8406	8396	2	3	5	6	8
33	·8387	8377	8368	8358	8348	8339	8329	8320	8310	8300	2	3	5	6	8
34	·8290	8281	8271	8261	8251	8241	8231	8221	8211	8202	2	3	5	7	8
35	·8192	8181	8171	8161	8151	8141	8131	8121	8111	8100	2	3	5	7	8
36	·8090	8080	8070	8059	8049	8039	8028	8018	8007	7997	2	3	5	7	9
37	·7986	7976	7965	7955	7944	7934	7923	7912	7902	7891	2	4	5	7	9
38	·7880	7869	7859	7848	7837	7826	7815	7804	7793	7782	2	4	5	7	9
39	·7771	7760	7749	7738	7727	7716	7705	7694	7683	7672	2	4	6	7	9
40°	·7660	7649	7638	7627	7615	7604	7593	7581	7570	7559	2	4	6	8	9
41	·7547	7536	7524	7513	7501	7490	7478	7466	7455	7443	2	4	6	8	10
42	·7431	7420	7408	7396	7385	7373	7361	7349	7337	7325	2	4	6	8	10
43	·7314	7302	7290	7278	7266	7254	7242	7230	7218	7206	2	4	6	8	10
44	·7193	7181	7169	7157	7145	7133	7120	7108	7096	7083	2	4	6	8	10

NATURAL COSINES

(N.B., Subtract Mean Differences)

	0′	6′	12′	18′	24′	30′	36′	42′	48′	54′	Mean Differences.				
											1′	2′	3′	4′	5′
45°	·7071	7059	7046	7034	7022	**7009**	6997	6984	6972	6959	2	4	6	8	10
46	·6947	6934	6921	6909	6896	**6884**	6871	6858	6845	6833	2	4	6	8	11
47	·6820	6807	6794	6782	6769	**6756**	6743	6730	6717	6704	2	4	6	9	11
48	·6691	6678	6665	6652	6639	**6626**	6613	6600	6587	6574	2	4	7	9	11
49	·6561	6547	6534	6521	6508	**6494**	6481	6468	6455	6441	2	4	7	9	11
50°	·6428	6414	6401	6388	6374	**6361**	6347	6334	6320	6307	2	4	7	9	11
51	·6293	6280	6266	6252	6239	**6225**	6211	6198	6184	6170	2	5	7	9	11
52	·6157	6143	6129	6115	6101	**6088**	6074	6060	6046	6032	2	5	7	9	12
53	·6018	6004	5990	5976	5962	**5948**	5934	5920	5906	5892	2	5	7	9	12
54	·5878	5864	5850	5835	5821	**5807**	5793	5779	5764	5750	2	5	7	9	12
55	·5736	5721	5707	5693	5678	**5664**	5650	5635	5621	5606	2	5	7	10	12
56	·5592	5577	5563	5548	5534	**5519**	5505	5490	5476	5461	2	5	7	10	12
57	·5446	5432	5417	5402	5388	**5373**	5358	5344	5329	5314	2	5	7	10	12
58	·5299	5284	5270	5255	5240	**5225**	5210	5195	5180	5165	2	5	7	10	12
59	·5150	5135	5120	5105	5090	**5075**	5060	5045	5030	5015	3	5	8	10	13
60°	·5000	4985	4970	4955	4939	**4924**	4909	4894	4879	4863	3	5	8	10	13
61	·4848	4833	4818	4802	4787	**4772**	4756	4741	4726	4710	3	5	8	10	13
62	·4695	4679	4664	4648	4633	**4617**	4602	4586	4571	4555	3	5	8	10	13
63	·4540	4524	4509	4493	4478	**4462**	4446	4431	4415	4399	3	5	8	10	13
64	·4384	4368	4352	4337	4321	**4305**	4289	4274	4258	4242	3	5	8	11	13
65	·4226	4210	4195	4179	4163	**4147**	4131	4115	4099	4083	3	5	8	11	13
66	·4067	4051	4035	4019	4003	**3987**	3971	3955	3939	3923	3	5	8	11	14
67	·3907	3891	3875	3859	3843	**3827**	3811	3795	3778	3762	3	5	8	11	14
68	·3746	3730	3714	3697	3681	**3665**	3649	3633	3616	3600	3	5	8	11	14
69	·3584	3567	3551	3535	3518	**3502**	3486	3469	3453	3437	3	5	8	11	14
70°	·3420	3404	3387	3371	3355	**3338**	3322	3305	3289	3°72	3	5	8	11	14
71	·3256	3239	3223	3206	3190	**3173**	3156	3140	3123	3107	3	6	8	11	14
72	·3090	3074	3057	3040	3024	**3007**	2990	2974	2957	2940	3	6	8	11	14
73	·2924	2907	2890	2874	2857	**2840**	2823	2807	2790	2773	3	6	8	11	14
74	·2756	2740	2723	2706	2689	**2672**	2656	2639	2622	2605	3	6	8	11	14
75	·2588	2571	2554	2538	2521	**2504**	2487	2470	2453	2436	3	6	8	11	14
76	·2419	2402	2385	2368	2351	**2334**	2317	2300	2284	2267	3	6	8	11	14
77	·2250	2233	2215	2198	2181	**2164**	2147	2130	2113	2096	3	6	9	11	14
78	·2079	2062	2045	2028	2011	**1994**	1977	1959	1942	1925	3	6	9	11	14
79	·1908	1891	1874	1857	1840	**1822**	1805	1788	1771	1582	3	6	9	11	14
80°	·1736	1719	1702	1685	1668	**1650**	1633	1616	1599	1582	3	6	9	12	14
81	·1564	1547	1530	1513	1495	**1478**	1461	1444	1426	1409	3	6	9	12	14
82	·1392	1374	1357	1340	1323	**1305**	1288	1271	1253	1236	3	6	9	12	14
83	·1219	1201	1184	1167	1149	**1132**	1115	1097	1080	1063	3	6	9	12	14
84	·1045	1028	1011	0993	0976	**0958**	0941	0924	0906	0889	3	6	9	12	14
85	·0872	0854	0837	0819	0802	**0785**	0767	0750	0732	0715	3	6	9	12	14
86	·0698	0680	0663	0645	0628	**0610**	0593	0576	0558	0541	3	6	9	12	15
87	·0523	0506	0488	0471	0454	**0436**	0419	0401	0384	0366	3	6	9	12	15
88	·0349	0332	0314	0297	0279	**0262**	0244	0227	0209	0192	3	6	9	12	15
89	·0175	0157	0140	0122	0105	**0087**	0070	0052	0035	0017	3	6	9	12	15

NATURAL TANGENTS

	0′	6′	12′	18′	24′	30′	36′	42′	48′	54′	Mean Differences.				
											1′	2′	3′	4′	5′
0°	·0000	0017	0035	0052	0070	**0087**	0105	0122	0140	0157	3	6	9	12	15
1	·0175	0192	0209	0227	0244	**0262**	0279	0297	0314	0332	3	6	9	12	15
2	·0349	0367	0384	0402	0419	**0437**	0454	0472	0489	0507	3	6	9	12	15
3	·0524	0542	0559	0577	0594	**0612**	0629	0647	0664	0682	3	6	9	12	15
4	·0699	0717	0734	0752	0769	**0787**	0805	0822	0840	0857	3	6	9	12	15
5	·0875	0892	0910	0928	0945	**0963**	0981	0998	1016	1033	3	6	9	12	15
6	·1051	1069	1086	1104	1122	**1139**	1157	1175	1192	1210	3	6	9	12	15
7	·1228	1246	1263	1281	1299	**1317**	1334	1352	1370	1388	3	6	9	12	15
8	·1405	1423	1441	1459	1477	**1495**	1512	1530	1548	1566	3	6	9	12	15
9	·1584	1602	1620	1638	1655	**1673**	1691	1709	1727	1745	3	6	9	12	15
10°	·1763	1781	1799	1817	1835	**1853**	1871	1890	1908	1926	3	6	9	12	15
11	·1944	1962	1980	1998	2016	**2035**	2053	2071	2089	2107	3	6	9	12	15
12	·2126	2144	2162	2180	2199	**2217**	2235	2254	2272	2290	3	6	9	12	15
13	·2309	2327	2345	2364	2382	**2401**	2419	2438	2456	2475	3	6	9	12	15
14	·2493	2512	2530	2549	2568	**2586**	2605	2623	2642	2661	3	6	9	12	16
15	·2679	2698	2717	2736	2754	**2773**	2792	2811	2830	2849	3	6	9	13	16
16	·2867	2886	2905	2924	2943	**2962**	2981	3000	3019	3038	3	6	9	13	16
17	·3057	3076	3096	3115	3134	**3153**	3172	3191	3211	3230	3	6	10	13	16
18	·3249	3269	3288	3307	3327	**3346**	3365	3385	3404	3424	3	6	10	13	16
19	·3443	3463	3482	3502	3522	**3541**	3561	3581	3600	3620	3	7	10	13	16
20°	·3640	3659	3679	3699	3719	**3739**	3759	3779	3799	3819	3	7	10	13	17
21	·3839	3859	3879	3899	3919	**3939**	3959	3979	4000	4020	3	7	10	13	17
22	·4040	4061	4081	4101	4122	**4142**	4163	4183	4204	4224	3	7	10	14	17
23	·4245	4265	4286	4307	4327	**4348**	4369	4390	4411	4431	3	7	10	14	17
24	·4452	4473	4494	4515	4536	**4557**	4578	4599	4621	4642	4	7	11	14	18
25	·4663	4684	4706	4727	4748	**4770**	4791	4813	4834	4856	4	7	11	14	18
26	·4877	4899	4921	4942	4964	**4986**	5008	5029	5051	5073	4	7	11	15	18
27	·5095	5117	5139	5161	5184	**5206**	5228	5250	5272	5295	4	7	11	15	18
28	·5317	5340	5362	5384	5407	**5430**	5452	5475	5498	5520	4	8	11	15	19
29	·5543	5566	5589	5612	5635	**5658**	5681	5704	5727	5750	4	8	12	15	19
30°	·5774	5797	5820	5844	5867	**5890**	5914	5938	5961	5985	4	8	12	16	20
31	·6009	6032	6056	6080	6104	**6128**	6152	6176	6200	6224	4	8	12	16	20
32	·6249	6273	6297	6322	6346	**6371**	6395	6420	6445	6469	4	8	12	16	20
33	·6494	6519	6544	6569	6594	**6619**	6644	6669	6694	6720	4	8	13	17	21
34	·6745	6771	6796	6822	6847	**6873**	6899	6924	6950	6976	4	9	13	17	21
35	·7002	7028	7054	7080	7107	**7133**	7159	7186	7212	7239	4	9	13	18	22
36	·7265	7292	7319	7346	7373	**7400**	7427	7454	7481	7508	5	9	14	18	23
37	·7536	7563	7590	7618	7646	**7673**	7701	7729	7757	7785	5	9	14	18	23
38	·7813	7841	7869	7898	7926	**7954**	7983	8012	8040	8069	5	9	14	19	24
39	·8098	8127	8156	8185	8214	**8243**	8273	8302	8332	8361	5	10	15	20	24
40°	·8391	8421	8451	8481	8511	**8541**	8571	8601	8632	8662	5	10	15	20	25
41	·8693	8724	8754	8785	8816	**8847**	8878	8910	8941	8972	5	10	16	21	26
42	·9004	9036	9067	9099	9131	**9163**	9195	9228	9260	9293	5	11	16	21	27
43	·9325	9358	9391	9424	9457	**9490**	9523	9556	9590	9623	6	11	17	22	28
44	·9657	9691	9725	9759	9793	**9827**	9861	9896	9930	9965	6	11	17	23	29

NATURAL TANGENTS

	0′	6′	12′	18′	24′	30′	36′	42′	48′	54′	1′	2′	3′	4′	5′
												Mean	Diffe	rences.	
45°	1·0000	0035	0070	0105	0141	0176	0212	0247	0283	0319	6	12	18	24	30
46	1·0355	0392	0428	0464	0501	0538	0575	0612	0649	0686	6	12	18	25	31
47	1·0724	0761	0799	0837	0875	0913	0951	0990	1028	1067	6	13	19	25	32
48	1·1106	1145	1184	1224	1263	1303	1343	1383	1423	1463	7	13	20	27	33
49	1·1504	1544	1585	1626	1667	1708	1750	1792	1833	1875	7	14	21	28	34
50°	1·1918	1960	2002	2045	2088	2131	2174	2218	2261	2305	7	14	22	29	36
51	1·2349	2393	2437	2482	2527	2572	2617	2662	2708	2753	8	15	23	30	38
52	1·2799	2846	2892	2938	2985	3032	3079	3127	3175	3222	8	16	24	31	39
53	1·3270	3319	3367	3416	3465	3514	3564	3613	3663	3713	8	16	25	33	41
54	1·3764	3814	3865	3916	3968	4019	4071	4124	4176	4229	9	17	26	34	43
55	1·4281	4335	4388	4442	4496	4550	4605	4659	4715	4770	9	18	27	36	45
56	1·4826	4882	4938	4994	5051	5108	5166	5224	5282	5340	10	19	29	38	48
57	1·5399	5458	5517	5577	5637	5697	5757	5818	5880	5941	10	20	30	40	50
58	1·6003	6066	6128	6191	6255	6319	6383	6447	6512	6577	11	21	32	43	53
59	1·6643	6709	6775	6842	6909	6977	7045	7113	7182	7251	11	23	34	45	56
60°	1·7321	7391	7461	7532	7603	7675	7747	7820	7893	7966	12	24	36	48	60
61	1·8040	8115	8190	8265	8341	8418	8495	8572	8650	8728	13	26	38	51	64
62	1·8807	8887	8967	9047	9128	9210	9292	9375	9458	9542	14	27	41	55	68
63	1·9626	9711	9797	9883	9970	0057	0145	0233	0323	0413	15	29	44	58	73
64	2·0503	0594	0686	0778	0872	0965	1060	1155	1251	1348	16	31	47	63	78
65	2·1445	1543	1642	1742	1842	1943	2045	2148	2251	2355	17	34	51	68	85
66	2·2460	2566	2673	2781	2889	2998	3109	3220	3332	3445	18	37	55	73	92
67	2·3559	3673	3789	3906	4023	4142	4262	4383	4504	4627	20	40	60	79	99
68	2·4751	4876	5002	5129	5257	5386	5517	5649	5782	5916	22	43	65	87	108
69	2·6051	6187	6325	6464	6605	6746	6889	7034	7179	7326	24	47	71	95	119
70°	2·7475	7625	7776	7929	8083	8239	8397	8556	8716	8878	26	52	78	104	131
71	2·9042	9208	9375	9544	9714	9887	0061	0237	0415	0595	29	58	87	116	145
72	3·0777	0961	1146	1334	1524	1716	1910	2106	2305	2506	32	64	96	129	161
73	3·2709	2914	3122	3332	3544	3759	3977	4197	4420	4646	36	72	108	144	180
74	3·4874	5105	5339	5576	5816	6059	6305	6554	6806	7062	41	81	122	163	204
75	3·7321	7583	7848	8118	8391	8667	8947	9232	9520	9812	46	93	139	186	232
76	4·0108	0408	0713	1022	1335	1653	1976	2303	2635	2972					
77	4·3315	3662	4015	4374	4737	5107	5483	5864	6252	6646					
78	4·7046	7453	7867	8288	8716	9152	9504	0045	0504	0970					
79	5·1446	1929	2422	2924	3435	3955	4486	5026	5578	6140					
80°	5·6713	7297	7894	8502	9124	9758	0405	1066	1742	2432					
81	6·3138	3859	4596	5350	6122	6912	7720	8548	9395	0264	*Mean* differences				
82	7·1154	2066	3002	3962	4947	5958	6996	8062	9158	0285	no longer suf-				
83	8·1443	2636	3863	5126	6427	7769	9152	0579	2052	3572	ficiently ac-				
84	9·514	9·677	9·845	10·02	10·20	10·39	10·58	10·78	10·99	11·20	curate.				
85	11·43	11·66	11·91	12·16	12·43	12·71	13·00	13·30	13·62	13·95					
86	14·30	14·67	15·06	15·46	15·89	16·35	16·83	17·34	17·89	18·46					
87	19·08	19·74	20·45	21·20	22·02	22·90	23·86	24·90	26·03	27·27					
88	28·64	30·14	31·82	33·69	35·80	38·19	40·92	44·07	47·74	52·08					
89	57·29	63·66	71·62	81·85	95·49	114·6	143·2	191·0	286·5	573·0					

RECIPROCALS

	0	1	2	3	4	5	6	7	8	9	1	2	3	4	5	6	7	8	9
											Subtract Differences.								
10	1000	9901	9804	9709	9615	9524	9434	9346	9259	9174									
11	9091	9009	8929	8850	8772	8696	8621	8547	8475	8403									
12	8333	8264	8197	8130	8065	8000	7937	7874	7813	7752			Mean differences						
13	7692	7634	7576	7519	7463	7407	7353	7299	7246	7194			not sufficiently						
14	7143	7092	7042	6993	6944	6897	6849	6803	6757	6711			accurate.						
15	6667	6623	6579	6536	6494	6452	6410	6369	6329	6289	4	8	13	17	21	25	29	33	38
16	6250	6211	6173	6135	6098	6061	6024	5988	5952	5917	4	7	11	15	18	22	26	29	33
17	5882	5848	5814	5780	5747	5714	5682	5650	5618	5587	3	6	10	13	16	20	23	26	29
18	5556	5525	5495	5464	5435	5405	5376	5348	5319	5291	3	6	9	12	15	17	20	23	26
19	5263	5236	5208	5181	5155	5128	5102	5076	5051	5025	3	5	8	11	13	16	18	21	24
20	5000	4975	4950	4926	4902	4878	4854	4831	4808	4785	2	5	7	10	12	14	17	19	21
21	4762	4739	4717	4695	4673	4651	4630	4608	4587	4566	2	4	7	9	11	13	15	17	19
22	4545	4525	4505	4484	4464	4444	4425	4405	4386	4367	2	4	6	8	10	12	14	16	18
23	4348	4329	4310	4292	4274	4255	4237	4219	4202	4184	2	4	5	7	9	11	13	14	16
24	4167	4149	4132	4115	4098	4082	4065	4049	4032	4016	2	3	5	7	8	10	12	13	15
25	4000	3984	3968	3953	3937	3922	3906	3891	3876	3861	2	3	5	6	8	9	11	12	14
26	3846	3831	3817	3802	3788	3774	3759	3745	3731	3717	1	3	4	6	7	8	10	11	13
27	3704	3690	3676	3663	3650	3636	3623	3610	3597	3584	1	3	4	5	7	8	9	11	12
28	3571	3559	3546	3534	3521	3509	3497	3484	3472	3460	1	2	4	5	6	7	9	10	11
29	3448	3436	3425	3413	3401	3390	3378	3367	3356	3344	1	2	3	5	6	7	8	9	10
30	3333	3322	3311	3300	3289	3279	3268	3257	3247	3236	1	2	3	4	5	6	7	9	10
31	3226	3215	3205	3195	3185	3175	3165	3155	3145	3135	1	2	3	4	5	6	7	8	9
32	3125	3115	3106	3096	3086	3077	3067	3058	3049	3040	1	2	3	4	5	6	7	8	9
33	3030	3021	3012	3003	2994	2985	2976	2967	2959	2950	1	2	3	4	4	5	6	7	8
34	2941	2933	2924	2915	2907	2899	2890	2882	2874	2865	1	2	3	3	4	5	6	7	8
35	2857	2849	2841	2833	2825	2817	2809	2801	2793	2786	1	2	2	3	4	5	6	6	7
36	2778	2770	2762	2755	2747	2740	2732	2725	2717	2710	1	2	2	3	4	5	5	6	7
37	2703	2695	2688	2681	2674	2667	2660	2653	2646	2639	1	1	2	3	4	4	5	6	6
38	2632	2625	2618	2611	2604	2597	2591	2584	2577	2571	1	1	2	3	3	4	5	5	6
39	2564	2558	2551	2545	2538	2532	2525	2519	2513	2506	1	1	2	3	3	4	4	5	6
40	2500	2494	2488	2481	2475	2469	2463	2457	2451	2445	1	1	2	2	3	4	4	5	5
41	2439	2433	2427	2421	2415	2410	2404	2398	2392	2387	1	1	2	2	3	3	4	5	5
42	2381	2375	2370	2364	2358	2353	2347	2342	2336	2331	1	1	2	2	3	3	4	4	5
43	2326	2320	2315	2309	2304	2299	2294	2288	2283	2278	1	1	2	2	3	3	4	4	5
44	2273	2268	2262	2257	2252	2247	2242	2237	2232	2227	1	1	2	2	3	3	4	4	5
45	2222	2217	2212	2208	2203	2198	2193	2188	2183	2179	0	1	1	2	2	3	3	4	4
46	2174	2169	2165	2160	2155	2151	2146	2141	2137	2132	0	1	1	2	2	3	3	4	4
47	2128	2123	2119	2114	2110	2105	2101	2096	2092	2088	0	1	1	2	2	3	3	4	4
48	2083	2079	2075	2070	2066	2062	2058	2053	2049	2045	0	1	1	2	2	3	3	3	4
49	2041	2037	2033	2028	2024	2020	2016	2012	2008	2004	0	1	1	2	2	2	3	3	4
50	2000	1996	1992	1988	1984	1980	1976	1972	1969	1965	0	1	1	2	2	2	3	3	4
51	1961	1957	1953	1949	1946	1942	1938	1934	1931	1927	0	1	1	2	2	2	3	3	3
52	1923	1919	1916	1912	1908	1905	1901	1898	1894	1890	0	1	1	1	2	2	3	3	3
53	1887	1883	1880	1876	1873	1869	1866	1862	1859	1855	0	1	1	1	2	2	2	3	3
54	1852	1848	1845	1842	1838	1835	1832	1828	1825	1821	0	1	1	1	2	2	2	3	3

RECIPROCALS

	0	1	2	3	4	5	6	7	8	9	Subtract Differences.								
											1	2	3	4	5	6	7	8	9
55	1818	1815	1812	1808	1805	**1802**	1799	1795	1792	1789	0	1	1	1	2	2	2	3	3
56	1786	1783	1779	1776	1773	**1770**	1767	1764	1761	1757	1	1	1	2	2	2	2	3	3
57	1754	1751	1748	1745	1742	**1739**	1736	1733	1730	1727	1	1	1	2	2	2	2	3	3
58	1724	1721	1718	1715	1712	**1709**	1706	1704	1701	1698	1	1	1	1	2	2	2	3	3
59	1695	1692	1689	1686	1684	**1681**	1678	1675	1672	1669	1	1	1	1	2	2	2	3	3
60	1667	1664	1661	1658	1656	**1653**	1650	1647	1645	1642	1	1	1	1	2	2	2	3	3
61	1639	1637	1634	1631	1629	**1626**	1623	1621	1618	1616	1	1	1	1	2	2	2	2	2
62	1613	1610	1608	1605	1603	**1600**	1597	1595	1592	1590	1	1	1	1	2	2	2	2	2
63	1587	1585	1582	1580	1577	**1575**	1572	1570	1567	1565	0	1	1	1	1	2	2	2	2
64	1563	1560	1558	1555	1553	**1550**	1548	1546	1543	1541	0	1	1	1	1	2	2	2	2
65	1538	1536	1534	1531	1529	**1527**	1524	1522	1520	1517	0	1	1	1	1	2	2	2	2
66	1515	1513	1511	1508	1506	**1504**	1502	1499	1497	1495	0	1	1	1	1	2	2	2	2
67	1493	1490	1488	1486	1484	**1481**	1479	1477	1475	1473	0	1	1	1	1	2	2	2	2
68	1471	1468	1466	1464	1462	**1460**	1458	1456	1453	1451	0	1	1	1	1	2	2	2	2
69	1449	1447	1445	1443	1441	**1439**	1437	1435	1433	1431	0	1	1	1	1	1	2	2	2
70	1429	1427	1425	1422	1420	**1418**	1416	1414	1412	1410	0	1	1	1	1	1	2	2	2
71	1408	1406	1404	1403	1401	**1399**	1397	1395	1393	1391	0	1	1	1	1	1	2	2	2
72	1389	1387	1385	1383	1381	**1379**	1377	1376	1374	1372	0	1	1	1	1	1	1	2	2
73	1370	1368	1366	1364	1362	**1361**	1359	1357	1355	1353	0	1	1	1	1	1	1	2	2
74	1351	1350	1348	1346	1344	**1342**	1340	1339	1337	1335	0	1	1	1	1	1	1	1	2
75	1333	1332	1330	1328	1326	**1325**	1323	1321	1319	1318	0	1	1	1	1	1	1	1	2
76	1316	1314	1312	1311	1309	**1307**	1305	1304	1302	1300	0	1	1	1	1	1	1	1	2
77	1299	1297	1295	1294	1292	**1290**	1289	1287	1285	1284	0	0	1	1	1	1	1	1	1
78	1282	1280	1279	1277	1276	**1274**	1272	1271	1269	1267	0	0	1	1	1	1	1	1	1
79	1266	1264	1263	1261	1259	**1258**	1256	1255	1253	1252	0	0	1	1	1	1	1	1	1
80	1250	1248	1247	1245	1244	**1242**	1241	1239	1238	1236	0	0	1	1	1	1	1	1	1
81	1235	1233	1232	1230	1229	**1227**	1225	1224	1222	1221	0	0	1	1	1	1	1	1	1
82	1220	1218	1217	1215	1214	**1212**	1211	1209	1208	1206	0	0	1	1	1	1	1	1	1
83	1205	1203	1202	1200	1199	**1198**	1196	1195	1193	1192	0	0	1	1	1	1	1	1	1
84	1190	1189	1188	1186	1185	**1183**	1182	1181	1179	1178	0	0	1	1	1	1	1	1	1
85	1176	1175	1174	1172	1171	**1170**	1168	1167	1166	1164	0	0	1	1	1	1	1	1	1
86	1163	1161	1160	1159	1157	**1156**	1155	1153	1152	1151	0	0	1	1	1	1	1	1	1
87	1149	1148	1147	1145	1144	**1143**	1142	1140	1139	1138	0	0	1	1	1	1	1	1	1
88	1136	1135	1134	1133	1131	**1130**	1129	1127	1126	1125	0	0	1	1	1	1	1	1	1
89	1124	1122	1121	1120	1119	**1117**	1116	1115	1114	1112	0	0	1	1	1	1	1	1	1
90	1111	1110	1109	1107	1106	**1105**	1104	1103	1101	1100	0	0	1	1	1	1	1	1	1
91	1099	1098	1096	1095	1094	**1093**	1092	1091	1089	1088	0	0	0	1	1	1	1	1	1
92	1087	1086	1085	1083	1082	**1081**	1080	1079	1078	1076	0	0	0	1	1	1	1	1	1
93	1075	1074	1073	1072	1071	**1070**	1068	1067	1066	1065	0	0	0	1	1	1	1	1	1
94	1064	1063	1062	1060	1059	**1058**	1057	1056	1055	1054	0	0	0	1	1	1	1	1	1
95	1053	1052	1050	1049	1048	**1047**	1046	1045	1044	1043	0	0	0	1	1	1	1	1	1
96	1042	1041	1040	1038	1037	**1036**	1035	1034	1033	1032	0	0	0	1	1	1	1	1	1
97	1031	1030	1029	1028	1027	**1026**	1025	1024	1022	1021	0	0	0	1	1	1	1	1	1
98	1020	1019	1018	1017	1016	**1015**	1014	1013	1012	1011	0	0	0	1	1	1	1	1	1
99	1010	1009	1008	1007	1006	**1005**	1004	1003	1002	1001	0	0	0	0	1	1	1	1	1

SQUARES

	0	1	2	3	4	5	6	7	8	9	Mean Differences.								
											1	2	3	4	5	6	7	8	9
1·0	1·000	1·020	1·040	1·061	1·082	**1·103**	1·124	1·145	1·166	1·188	2	4	6	8	**10**	13	15	17	19
1·1	1·210	1·232	1·254	1·277	1·300	**1·323**	1·346	1·369	1·392	1·416	2	5	7	9	**11**	14	16	18	21
1·2	1·440	1·464	1·488	1·513	1·538	**1·563**	1·588	1·613	1·638	1·664	2	5	7	10	**12**	15	17	20	22
1·3	1·690	1·716	1·742	1·769	1·796	**1·823**	1·850	1·877	1·904	1·932	3	5	8	11	**13**	16	19	22	24
1·4	1·960	1·988	2·016	2·045	2·074	**2·103**	2·132	2·161	2·190	2·220	3	6	9	12	**14**	17	20	23	26
1·5	2·250	2·280	2·310	2·341	2·372	**2·403**	2·434	2·465	2·496	2·528	3	6	9	12	**15**	19	22	25	28
1·6	2·560	2·592	2·624	2·657	2·690	**2·723**	2·756	2·789	2·822	2·856	3	7	10	13	**16**	20	23	26	30
1·7	2·890	2·924	2·958	2·993	3·028	**3·063**	3·098	3·133	3·168	3·204	3	7	10	14	**17**	21	24	28	31
1·8	3·240	3·276	3·312	3·349	3·386	**3·423**	3·460	3·497	3·534	3·572	4	7	11	15	**18**	22	26	30	33
1·9	3·610	3·648	3·686	3·725	3·764	**3·803**	3·842	3·881	3·920	3·960	4	8	12	16	**19**	23	27	31	35
2·0	4·000	4·040	4·080	4·121	4·162	**4·203**	4·244	4·285	4·326	4·368	4	8	12	16	**20**	25	29	33	37
2·1	4·410	4·452	4·494	4·537	4·580	**4·623**	4·666	4·709	4·752	4·796	4	9	13	17	**21**	26	30	34	39
2·2	4·840	4·884	4·928	4·973	5·018	**5·063**	5·108	5·153	5·198	5·244	4	9	13	18	**22**	27	31	36	40
2·3	5·290	5·336	5·382	5·429	5·476	**5·523**	5·570	5·617	5·664	5·712	5	9	14	19	**23**	28	33	38	42
2·4	5·760	5·808	5·856	5·905	5·954	**6·003**	6·052	6·101	6·150	6·200	5	10	15	20	**24**	29	34	39	44
2·5	6·250	6·300	6·350	6·401	6·452	**6·503**	6·554	6·605	6·656	6·708	5	10	15	20	**25**	31	36	41	46
2·6	6·760	6·812	6·864	6·917	6·970	**7·023**	7·076	7·129	7·182	7·236	5	11	16	21	**26**	32	37	42	48
2·7	7·290	7·344	7·398	7·453	7·508	**7·563**	7·618	7·673	7·728	7·784	5	11	16	22	**27**	33	38	44	49
2·8	7·840	7·896	7·952	8·009	8·066	**8·123**	8·180	8·237	8·294	8·352	6	11	17	23	**28**	34	40	46	51
2·9	8·410	8·468	8·526	8·585	8·644	**8·703**	8·762	8·821	8·880	8·940	6	12	18	24	**29**	35	41	47	53
3·0	9·000	9·060	9·120	9·181	9·242	**9·303**	9·364	9·425	9·486	9·548	6	12	18	24	**30**	37	43	49	55
3·1 {	9·610	9·672	9·734	9·797	9·860	**9·923**	9·986				6	13	19	25	**31**	38	44	50	57
								10·05	10·11	10·18	1	1	2	3	3	4	5	5	6
3·2	10·24	10·30	10·37	10·43	10·50	**10·56**	10·63	10·69	10·76	10·82	1	1	2	3	3	4	5	5	6
3·3	10·89	10·96	11·02	11·09	11·16	**11·22**	11·29	11·36	11·42	11·49	1	1	2	3	3	4	5	5	6
3·4	11·56	11·63	11·70	11·76	11·83	**11·90**	11·97	12·04	12·11	12·18	1	1	2	3	3	4	5	6	6
3·5	12·25	12·32	12·39	12·46	12·53	**12·60**	12·67	12·74	12·82	12·89	1	1	2	3	4	4	5	6	6
3·6	12·96	13·03	13·10	13·18	13·25	**13·32**	13·40	13·47	13·54	13·62	1	1	2	3	4	4	5	6	7
3·7	13·69	13·76	13·84	13·91	13·99	**14·06**	14·14	14·21	14·29	14·36	1	2	2	3	4	4	5	6	7
3·8	14·44	14·52	14·59	14·67	14·75	**14·82**	14·90	14·98	15·05	15·13	1	2	2	3	4	5	5	6	7
3·9	15·21	15·29	15·37	15·44	15·52	**15·60**	15·68	15·76	15·84	15·92	1	2	2	3	4	5	5	6	7
4·0	16·00	16·08	16·16	16·24	16·32	**16·40**	16·48	16·56	16·65	16·73	1	2	2	3	4	5	6	6	7
4·1	16·81	16·89	16·97	17·06	17·14	**17·22**	17·31	17·39	17·47	17·56	1	2	2	3	4	5	6	7	7
4·2	17·64	17·72	17·81	17·89	17·98	**18·06**	18·15	18·23	18·32	18·40	1	2	3	3	4	5	6	7	8
4·3	18·49	18·58	18·66	18·75	18·84	**18·92**	19·01	19·10	19·18	19·27	1	2	3	3	4	5	6	7	8
4·4	19·36	19·45	19·54	19·62	19·71	**19·80**	19·89	19·98	20·07	20·16	1	2	3	4	4	5	6	7	8
4·5	20·25	20·34	20·43	20·52	20·61	**20·70**	20·79	20·88	20·98	21·07	1	2	3	4	5	5	6	7	8
4·6	21·16	21·25	21·34	21·44	21·53	**21·62**	21·72	21·81	21·90	22·00	1	2	3	4	5	6	7	7	8
4·7	22·09	22·18	22·28	22·37	22·47	**22·56**	22·66	22·75	22·85	22·94	1	2	3	4	5	6	7	8	9
4·8	23·04	23·14	23·23	23·33	23·43	**23·52**	23·62	23·72	23·81	23·91	1	2	3	4	5	6	7	8	9
4·9	24·01	24·11	24·21	24·30	24·40	**24·50**	24·60	24·70	24·80	24·90	1	2	3	4	5	6	7	8	9
5·0	25·00	25·10	25·20	25·30	25·40	**25·50**	25·60	25·70	25·81	25·91	1	2	3	4	5	6	7	8	9
5·1	26·01	26·11	26·21	26·32	26·42	**26·52**	26·63	26·73	26·83	26·94	1	2	3	4	5	6	7	8	9
5·2	27·04	27·14	27·25	27·35	27·46	**27·56**	27·67	27·77	27·88	27·98	1	2	3	4	5	6	7	8	9
5·3	28·09	28·20	28·30	28·41	28·52	**28·62**	28·73	28·84	28·94	29·05	1	2	3	4	5	6	7	9	10
5·4	29·16	29·27	29·38	29·48	29·59	**29·70**	29·81	29·92	30·03	30·14	1	2	3	4	5	7	8	9	10

	0	1	2	3	4	5	6	7	8	9	Mean Differences.								
											1	2	3	4	5	6	7	8	9
5·5	30·25	30·36	30·47	30·58	30·69	**30·80**	30·91	31·02	31·14	31·25	1	2	3	4	6	7	8	9	10
5·6	31·36	31·47	31·58	31·70	31·81	**31·92**	32·04	32·15	32·26	32·38	1	2	3	5	6	7	8	9	10
5·7	32·49	32·60	32·72	32·83	32·95	**33·06**	33·18	33·29	33·41	33·52	1	2	3	5	6	7	8	9	10
5·8	33·64	33·76	33·87	33·99	34·11	**34·22**	34·34	34·46	34·57	34·69	1	2	4	5	6	7	8	9	11
5·9	34·81	34·93	35·05	35·16	35·28	**35·40**	35·52	35·64	35·76	35·88	1	2	4	5	6	7	8	10	11
6·0	36·00	36·12	36·24	36·36	36·48	**36·60**	36·72	36·84	36·97	37·09	1	2	4	5	6	7	8	10	11
6·1	37·21	37·33	37·45	37·58	37·70	**37·82**	37·95	38·07	38·19	38·32	1	2	4	5	6	7	9	10	11
6·2	38·44	38·56	38·69	38·81	38·94	**39·06**	39·19	39·31	39·44	39·56	1	3	4	5	6	8	9	10	11
6·3	39·69	39·82	39·94	40·07	40·20	**40·32**	40·45	40·58	40·70	40·83	1	3	4	5	6	8	9	10	11
6·4	40·96	41·09	41·22	41·34	41·47	**41·60**	41·73	41·86	41·99	42·12	1	3	4	5	6	8	9	10	12
6·5	42·25	42·38	42·51	42·64	42·77	**42·90**	43·03	43·16	43·30	43·43	1	3	4	5	7	8	9	10	12
6·6	43·56	43·69	43·82	43·96	44·09	**44·22**	44·36	44·49	44·62	44·76	1	3	4	5	7	8	9	11	12
6·7	44·89	45·02	45·16	45·29	45·43	**45·56**	45·70	45·83	45·97	46·10	1	3	4	5	7	8	9	11	12
6·8	46·24	46·38	46·51	46·65	46·79	**46·92**	47·06	47·20	47·33	47·47	1	3	4	5	7	8	10	11	12
6·9	47·61	47·75	47·89	48·02	48·16	**48·30**	48·44	48·58	48·72	48·86	1	3	4	6	7	8	10	11	13
7·0	49·00	49·14	49·28	49·42	49·56	**49·70**	49·84	49·98	50·13	50·27	1	3	4	6	7	8	10	11	13
7·1	50·41	50·55	50·69	50·84	50·98	**51·12**	51·27	51·41	51·55	51·70	1	3	4	6	7	9	10	11	13
7·2	51·84	51·98	52·13	52·27	52·42	**52·56**	52·71	52·85	53·00	53·14	1	3	4	6	7	9	10	12	13
7·3	53·29	53·44	53·58	53·73	53·88	**54·02**	54·17	54·32	54·46	54·61	1	3	4	6	7	9	10	12	13
7·4	54·76	54·91	55·06	55·20	55·35	**55·50**	55·65	55·80	55·95	56·10	1	3	4	6	7	9	10	12	13
7·5	56·25	56·40	56·55	56·70	56·85	**57·00**	57·15	57·30	57·46	57·61	2	3	5	6	8	9	11	12	14
7·6	57·76	57·91	58·06	58·22	58·37	**58·52**	58·68	58·83	58·98	59·14	2	3	5	6	8	9	11	12	14
7·7	59·29	59·44	59·60	59·75	59·91	**60·06**	60·22	60·37	60·53	60·68	2	3	5	6	8	9	11	13	14
7·8	60·84	61·00	61·15	61·31	61·47	**61·62**	61·78	61·94	62·09	62·25	2	3	5	6	8	9	11	13	14
7·9	62·41	62·57	62·73	62·88	63·04	**63·20**	63·36	63·52	63·68	63·84	2	3	5	6	8	10	11	13	14
8·0	64·00	64·16	64·32	64·48	64·64	**64·80**	64·96	65·12	65·29	65·45	2	3	5	6	8	10	11	13	14
8·1	65·61	65·77	65·93	66·10	66·26	**66·42**	66·59	66·75	66·91	67·08	2	3	5	7	8	10	11	13	15
8·2	67·24	67·40	67·57	67·73	67·90	**68·06**	68·23	68·39	68·56	68·72	2	3	5	7	8	10	12	13	15
8·3	68·89	69·06	69·22	69·39	69·56	**69·72**	69·89	70·06	70·22	70·39	2	3	5	7	8	10	12	13	15
8·4	70·56	70·73	70·90	71·06	71·23	**71·40**	71·57	71·74	71·91	72·08	2	3	5	7	8	10	12	13	15
8·5	72·25	72·42	72·59	72·76	72·93	**73·10**	73·27	73·44	73·62	73·79	2	3	5	7	9	10	12	14	15
8·6	73·96	74·13	74·30	74·48	74·65	**74·82**	75·00	75·17	75·34	75·52	2	3	5	7	9	10	12	14	16
8·7	75·69	75·86	76·04	76·21	76·39	**76·56**	76·74	76·91	77·09	77·26	2	4	5	7	9	11	12	14	16
8·8	77·44	77·62	77·79	77·97	78·15	**78·32**	78·50	78·68	78·85	79·03	2	4	5	7	9	11	13	14	16
8·9	79·21	79·39	79·57	79·74	79·92	**80·10**	80·28	80·46	80·64	80·82	2	4	5	7	9	11	13	14	16
9·0	81·00	81·18	81·36	81·54	81·72	**81·90**	82·08	82·26	82·45	82·63	2	4	5	7	9	11	13	14	16
9·1	82·81	82·99	83·17	83·36	83·54	**83·72**	83·91	84·09	84·27	84·46	2	4	5	7	9	11	13	15	16
9·2	84·64	84·82	85·01	85·19	85·38	**85·56**	85·75	85·93	86·12	86·30	2	4	6	7	9	11	13	15	17
9·3	86·49	86·68	86·86	87·05	87·24	**87·42**	87·61	87·80	87·98	88·17	2	4	6	7	9	11	13	15	17
9·4	88·36	88·55	88·74	88·92	89·11	**89·30**	89·49	89·68	89·87	90·06	2	4	6	8	9	11	13	15	17
9·5	90·25	90·44	90·63	90·82	91·01	**91·20**	91·39	91·58	91·78	91·97	2	4	6	8	10	11	13	15	17
9·6	92·16	92·35	92·54	92·74	92·93	**93·12**	93·32	93·51	93·70	93·90	2	4	6	8	10	12	14	15	17
9·7	94·09	94·28	94·48	94·67	94·87	**95·06**	95·26	95·45	95·65	95·84	2	4	6	8	10	12	14	16	18
9·8	96·04	96·24	96·43	96·63	96·83	**97·02**	97·22	97·42	97·61	97·81	2	4	6	8	10	12	14	16	18
9·9	98·01	98·21	98·41	98·60	98·80	**99·00**	99·20	99·40	99·60	99·80	2	4	6	8	10	12	14	16	18